W0234940

Ethics in Human-like Robots

Kamil Mamak
Department of Practical Philosophy, RADAR
University of Helsinki, Finland
and
Department of Criminal Law
Jagiellonian University, Kraków, Poland

A SCIENCE PUBLISHERS BOOK

First edition published 2025
by Routledge
2385 NW Executive Center Drive, Suite 320, Boca Raton FL 33431

and by Routledge
4 Park Square, Milton Park, Abingdon, Oxon, OX14 4RN

Routledge imprint an of Taylor & Francis Group, LLC

Library of Congress Cataloging-in-Publication Data (applied for)

ISBN: 978-1-032-65623-6 (hbk)
ISBN: 978-1-032-67104-8 (pbk)
ISBN: 978-1-032-67102-4 (ebk)
DOI: 10.1201/9781032671024

Typeset by Prime Publishing Services

Acknowledgments

I would like to start by thanking Pekka Mäkelä and Raul Hakli for creating a peaceful environment for intellectual work which is the research group RADAR: Robophilosophy, AI Ethics, and Datafication, at the Department of Practical Philosophy, University of Helsinki. I would like to express my gratitude to the other members of our research group, Dane Leigh Gogoshin, Tomi Kokkonen, Olli Niinivaara, Pii Telakivi, and Tuomas Vesterinen.

I had an opportunity to present ideas from this book at various academic events. I presented the outline of this book at the "HumanTech Meetings II: From Myths and Legends to Robots. Ethics in Designing Artificial Beings," organized at the SWPS University, Warsaw, Poland. I also presented at scientific conferences, including the "Design Matters. Conference on Emerging Technologies and Behavioural-driven Designs" at the Erasmus University Rotterdam, Netherlands; the 2023 Asian Law Schools Association (ALSA) Conference at the National Yang Ming Chiao Tung University, Taiwan; the conference "Phenomenology of Human–Technology Relation: A New Perspective on Technological Design," at the University of Bremen, Germany, the workshop "Philosophy of Avatars", at the University of Aberdeen, Scotland, UK, and the conference "Technology and Sustainable Development 2023 (TSD23)" at the Østfold University College, Halden, Norway.

I am grateful to my brother, Bartosz Mamak, who created the illustrations for this book.

I want to thank my wonderful wife Sylwia and son Leon.

Research for this book was financed by the Research Council of Finland (decision number: 333873).

Contents

5. Human Likeness of Robots and Moral Patiency

6. Conclusion: Towards Ethical Design of Human-like Robots

1 | Introduction

Introduction

The idea of creating artificial humans has a long history in human culture. Ancient myths narrate instances of artificial humans brought to life by gods (Mayor 2018). The word "robot" originates from a 1920 play that was about artificial humans made from artificial flesh who aimed to serve real humans (Čapek 2015). Due to advancements in robotics, the materialization of this idea is more real than ever. We are witnessing attempts to create humanoid robots that might be deployed in many spheres of our lives—police, healthcare, even love and sex. This book focuses on the ethical issues resulting from the human likeness of robots. With their bodies operating in the physical world, human-like robots bring ideas and risks that need to be addressed before their broader deployment.

But is it true that we are close to human-like robots? In the *Financial Times*, we can read about the "humanoid robot wave" (Huber 2024). Let us look at some recent news stories from the robotic world. In April 2024, Tesla announced that it would start selling its first humanoid robot, called Optimus, by the end of next year (Sriram 2024). Also in April 2024, Boston Dynamics, the robotic company known mostly for their robot dog Spot and humanoid robot Atlas, presented a new generation of the latter, which is fully electric, replacing the old hydraulic version (Roth 2024). Another news item from April 2024 concerns Sanctuary AI announcing the next generation of their humanoid robot, Phoenix. The chief executive officer (CEO) of the company said, "With generation seven, we have a system that we believe is the most closely analogous to a person of any available" (Heater 2024). In November 2023, we read that China plans to mass produce advanced humanoid robots by 2025 (Mann 2023). Around the same time, information was circulated about the opening of a factory in the US by Agility Robotics, which plans to mass produce Digit, their humanoid robot (Kolodny 2023). In a *New York Times* article commenting on the opening of the factory of Agility Robotics, there is a sentence: "the long-anticipated robot revolution has begun" (Howe and Antaya 2023).

As we can see from the selection of the recent news, humanoid robots might soon populate our world. At the same time, the ethical literature about human-like robots is full of concerns and underlying risks of having such robots around. For example, Alsegier points out that human-like robots are "one of the most controversial facets of modern technology" (Alsegier 2016: 24). Russell notes that "there is no good reason for robots to have humanoid form. There are also good, practical reasons not to have humanoid form" (Russell 2019: 126). Darling, in her book about robots, points out, "The main problem of anthropomorphism in robotics is that, right now, we aren't treating it as a matter of contention" (Darling 2021: 155). She believes that there is not enough discussion about this and that we are deploying robots without fully understanding the impact of anthropomorphism on people. The more human-like the robot is, the easier it is to anthropomorphize it (Gasser 2021: 334).

The fact that we have already got used to robotic vacuum cleaners, smart speakers, and delivery robots does not mean that the natural step is to accept human-like robots. Humanoid robots, with their human likeness, bring additional ethically relevant issues that should be discussed first. In this book, I focus on the ethically qualitative shift in designing robots that resemble humans. Throughout the book many questions will be asked that relate to ethical issues of human likeness; among these are: Is it safe to have human-like robots around us? Who would human-like robots represent, and why should it be a matter of concern? To what extent are human robots achievable? Is it ethical to have too human-like robots? Could robots have human-like ethics? Could robots be responsible in a human-like way? How should we treat robots that look like us? How should we treat robots that are like us? How do we mitigate the risks resulting from human-like robots? All these questions will be covered in the chapters that follow.

In recent years, numerous books with a focus on the ethical aspects of robots have been published. Besides the already mentioned book by Kate Darling, there are many other great books published (e.g., Gunkel 2018, 2023; Coeckelbergh 2022; Nyholm 2020; Wynsberghe 2016; Sweeney 2023; Gellers 2020; Tzafestas 2015; Devlin 2018; Turner 2019). There are also numerous journal papers, collective volumes, and conference proceedings, for example, from the Robophilsophy conference series (see, e.g., Hakli et al. 2023). Is there space for another book on the ethics of robots? There are already many critical voices about human-like robots in the literature. However, it seems that the academic discussion does not resonate enough with the discussion by the general public. This book aims to contribute to both discussions by adding new insights to scholarly discourse on the ethics of robots and by drawing attention specifically to human likeness in robots. The book deliberately narrows down to this one characteristic of robots—human likeness—to emphasize it as a standalone ethical problem. The growing number of human-like robots in real life and the significance of the potential consequences of their deployment indicate that human-like robots have the potential to be socially disruptive soon. It is, therefore, also an aspect of timing. It is now that human likeness in robots must be discussed before some risks become a reality.

Key Concepts and Ideas

Concepts like "robot," "ethics," "humanoid," "design," "human-like," and "anthropomorphism" have already been mentioned. Now, I unpack these concepts, which allows me to explain key ideas behind the book's scheme. Unpacking will be divided into three groups of matters. First, I will focus on robots; second, on ethics; and third, on design.

What is a Human-like Robot?

Robot

It is not surprising to say that there is no single definition of a robot. Coeckelbergh notes that there is no agreement on the definition of a robot, neither among roboticists nor among philosophers (Coeckelbergh 2022: 7). The word "robot" originated from the 1920 Czech play by Karl Čapek (Čapek 2015), in which robots were made from artificial flesh, which is much further away from the contemporary visions of typical robots. In literature, there are many takes on what a robot is. For example, Winfield proposes three definitions, and the most synthetic one defines a robot as "an embodied artificial intelligence" (Winfield 2012: 8). The popular frame that is used to define robots is the "sense–think–act"/"sense–plan–act" paradigm that underlines three components (see, e.g., Bartneck et al. 2021). First, a robot needs to have the ability to collect data from the external world. Second, it needs to have the ability to process it; this is the "thinking"/"planning" part. Finally, it needs to have the ability to influence the world. Jordan discusses the difficulties in defining robots and wonders what reasons there might be for concerns in this regard. He notes that it is hard to say what a robot is because the technology is constantly evolving, as well as the social context in which robots might be deployed. He also mentions the role of science fiction that influences the way in which we think about robots (Jordan 2016). For the purposes of this book, there is no need to spend more time discussing the general definition of a robot. The focus here is not on robots in a general sense, but on specific kinds of robots that are human like. Let us turn now to this characteristic.

Human-like Robot

The title of this book uses the term "human-like" to characterize the robots that are the subject of study. The candidate for substituting this term was the word "humanoid." This book uses both terms interchangeably. In general, this is about robots that resemble humans. What those terms mean in detail is more complex and requires more explanation. The resemblance might concern both the external aspects, like the human form, as well as aspects that are more internal. According to one of the definitions from the *Oxford English Dictionary*, humanoid means "with human form; having human characteristics." The human likeness is understood here in a much broader manner than mere physical similarity.

The other important notion for further deliberations is that human likeness comes in degrees. In a different context, it could mean something else. For example, we will talk about robots that can appear human like from a distance by having equivalents of a head, two legs, and two arms and being the size of humans, but are nevertheless from a close distance impossible to mistake for humans. On the other spectrum of human likeness are robots which are indistinguishable from humans at the visual level or possess some important human-like properties. Ethical risks are related to the whole spectrum of human likeness, and it is in this context that this book underlines that the design choice of any resemblance to humans is ethically marked.

Human-like robots could get close to humans, but never will be humans in all senses, which is also ethically problematic to some extent. To be clear, there are good reasons not to recreate humans fully in robots. What is more, it seems impossible in many aspects. However, if we decide to create human-shaped robots, we need to know that it is, to some extent, a false promise that might cause confusion and be disruptive to human values. Humans, in general, assume that human-like entities behave like humans, care about things that we care about, are aware of the consequences of their actions, have a self-preservation instinct, and so on. If human-like entities are just robots, these human characteristics do not apply. Human-like entities that do not behave as we would expect humans to behave might lead to risks for humans interacting with such robots. Later in this book, issues related to human-like ethics or human-like responsibility will be raised.

At the beginning of this chapter, it was mentioned that human-like robots might be the realization of the long-awaited dream of creating artificial humans. In the *Springer Handbook of Robotics*, in the entry chapter to the section devoted to humans and robots, Fitzpatrick et al. (2016) explain in a more systematic way why we build human-like robots. They present these reasons under six subsections. First, humans who are builders take their own abilities and characteristics as the most natural point of reference that is available to everyone. The second reason is that humans are a favorite subject for humans. We like to be around humans, as well as children in their early stages of development and as fully mature beings that constantly talk about other human beings. Third, researchers also think that the human form of robotics allows us to understand human intelligence better by mimicking it in machines' human properties. Humanoid robots could allow us to have a better opportunity to understand ourselves. Fourth, our world is shaped by humans for humans. Doors, stairs, and our technologies are accustomed to serving humans with certain biological characteristics, and this environment sets boundaries for robots. If we want to have a robot that will take groceries from the car, carry them on the stairs, put them in the fridge, prepare dinner, and wash the dishes, then all these tasks combine to limit the design choices of such a robot and make it closer to the human shape. Fifth, we are used to human interactions, and making robots human-like means it is easier to introduce robots to humans. Finally, the human form seems to be unavoidable for some of the applications of robots, like providing entertainment for adults by companion robots (Fitzpatrick et al. 2016: 1790–92).

Anthropomorphism

Human likeness is strongly linked with the term "anthropomorphism." In short, anthropomorphism is the human tendency to attribute human-like characteristics to other entities and events (see, e.g., Guthrie 1997). As Złotowski et al. note, humans anthropomorphize environments, animals, events, and robots, but robots are unique compared to other anthropomorphized events and objects because of their "higher anthropomorphic ability" (Złotowski et al. 2015: 347). According to Perugia et al., when human-like qualities are designed and included in the process by the designer, we can describe the anthropomorphic encoding.

When people attribute human-like qualities to robots, then it could be called anthropomorphic decoding (Perugia et al. 2022: 111). Seeing human likeness in robots could be the result of intentional choices, as well as unintended side effects of designing decisions. As Gasser points out, we have the tendency to see the world through an anthropomorphic filter; we use knowledge about ourselves as the basis of interpretation and for predicting the behavior of unfamiliar entities that we approach, and this tendency increases with the human likeness of the entity (Gasser 2021: 333–34).

Using anthropomorphic design in robots, which could mean the robot's shape, behavior, and interaction/communication with the human (Fink 2012: 200), could boost performance in robot interactions (Spatola et al. 2020). So, it could be assumed that in general, the more human-like the robot is, the better/smoother/easier interactions with robots will be. In that context, it is worth mentioning the "uncanny valley" theory from 1970 by Mashiro Mori (Mori 1970). In short, according to this theory, the more human-like a robot is, the more positive responses it gets, but this curve breaks down when robots are too human-like. It is pointed out that there is no empirical evidence that could support this hypothesis (Zlotowski et al. 2013). Also, a more recent review paper on the uncanny valley effect noted that "no conclusions regarding proposed adverse effects at higher levels of human likeness can be made at this stage" (Mara et al. 2022: 33).

The tendency to anthropomorphize robots is sometimes criticized and presented as something inappropriate. But in response to such criticism, Gunkel points out that "anthropomorphism is not a bug to be eliminated; it is a feature" (Gunkel 2023: 45). His works especially focus on the way how humans treat robots and do not agree with the voices that robots do not deserve to be treated better than mere things due to their ontological properties. There will be more on the aspects of human-like robots later in the book.

What is the Ethics of Human-like Robots?

Ethics

The other main concept in the title of this book is ethics. In general, ethics deal with questions such as how we ought to live, what is good, what is wrong, and what principles should we follow (for an overview, see, e.g., Gensler 2011; Shafer-Landau 2020). This book is interested in these questions in the context of robots. Writing a book about ethical issues could take various forms. In this book, I have followed

Coeckelbergh and his approach in his book *Robot Ethics*, which focuses on the problems that arise regarding robots, and not on ethical theories as such (Coeckelbergh 2022: 7). Some ethical theories are mentioned in this book, but more as background for specific problems.

One might notice that there are not yet many problems with human-like robots, simply because their presence so far is limited. In this context, the Collingridge dilemma should be mentioned (Collingridge 1981; see also Kudina and Verbeek 2019). According to this concept, technologies are relatively easy to regulate at the early stage of their development, but we do not yet know their societal consequences. However, when we do know their societal impact, it is often too late to impose effective regulations. In the last few decades, we have experienced many technologies that have had disruptive impacts on societies, like social media platforms or smartphones, and now, when those technologies are already around us, the regulatory efforts have limited impact.

Going back to robots, they are not yet present on a scale that is comparable to smartphones; but when they are around, it might be too late to be adverse to negative changes that they would have caused. The way to limit the potential negative consequences is to try to anticipate them (see Brey 2012). De Pagter notes that "many of the hopes, fears and expectations around robotics are related to anticipatory notions that engage with robots' potential societal ramification and include a wide range of uncertainties with regard to their (future) impact" (de Pagter 2023: 2006). He adds that there is a risk that anticipating scenarios could go wrong (see, e.g., de Pagter 2023: 2015). However, if we want to avoid at least some problems, then it seems that we need to come to terms with the possibility of being wrong. In his book *AI Ethics*, Coeckelbergh states that there is a need to discuss ethical issues "before it is too late" (Coeckelbergh 2020: 10).

This book focuses mostly on robots that are here and now or are technologically possible. However, in some contexts, there is a discussion concerning the possible characteristics of robots. I discuss them to point out the ethical risks of taking steps to build such robots.

Roboethics

The ethical issues that are raised in this book are related to robots. Such problems are sometimes discussed under the name "roboethics." Roboethics is a relatively new term, dating from 20 years ago. Veruggio pointed out that the beginning of a branch of ethics devoted to robots was related to the First International Symposium on Roboethics in 2004 (Veruggio 2005). Before that, the word roboethics did not exist in encyclopedias or on the Internet (Veruggio et al. 2016: 2136). However, this does not mean that the ethical issues of robotics were not discussed before; robots as subject of pop culture were discussed in the context of different, ethically related issues. Also, scholars worked on something that currently would be classified as robotics; for example, Putman, in his paper "Robots: Machines or Artificially Created Life?" was wondering about the possible moral status of machines (Putman 1964).

Veruggio, in the paper "The Birth of Roboethics," justifies the societal importance of the emerging subfield of ethics, and identifies a couple of fields of its application: the economy (in focusing on the employment consequences of robotization), effects on society (how robot servants change humans who can be addicted to them), healthcare, lack of access (inequalities in distribution of robots in society), deliberate abuse/terrorism (impact on war), and law (the issue of who is responsible for the actions of robots) (Veruggio 2005). In the paper, "What Should We Want From a Robot Ethic?" Asaro lists three things that we might mean by "ethics in robotics": "the ethical systems built into robots, the ethics of people who design and use robots, and the ethics of how people treat robots" (Asaro 2006: 9). In his opinion, the emerging subfield of ethics should combine all perspectives into one. He also underlines that robots should not be reduced to technical aspects, and there is a need to consider their societal relevance.

Lin, in his introductory chapter "Introduction to Robot Ethics," maps ethical and societal issues into three main clusters: "Safety and Errors" (harm to humans, issues of hacking, with examples from military fields); "Law and Ethics" (with a focus on the responsibility issues, privacy, machine ethics, relationships with humans, the moral status of robots); and finally "Social Impact", when he mentions issues such as the job losses as a result of automation, human dependency on robots, and the impact of robots on relationships and environment (Lin 2012). As we can see, there are several ways of categorizing ethical problems related to robots. Human likeness is an additional perspective that might change the accent of roboethics.

Ethics of Human-like Robots

There are three categories of relationships between issues of general roboethics and the ethics of human-like robots:

- ethical issues that concern human-like robots as well as other robots
- ethical issues that are modified by the human likeness of robots
- ethical issues that are unique to human-like robots

The first category concerns the issues that are more or less the same for humanoid robots and nonhuman-like robots. To some extent, this category pertains to questions concerning the safety of robots or privacy issues, autonomy, and responsibility, to name a few. For example, cameras in robotic vacuum cleaners as well as cameras that are parts of human-like robots allow for people to be observed. Some ethical issues could be worsened in the case of humanoid robots. Łichocki et al. mention that human likeness is the factor that could amplify ethical problems (Łichocki et al. 2011). For example, mentioned before privacy, Calo points out that sex robots could create an opportunity to collect intimate data that would not be possible to collect from an industrial robot (Calo 2012; see also Dodig-Crnkovic et al. 2021: 95). There are also categories of ethical issues that arise uniquely from the human likeness of robots. Examples from the last category are the issues with representation (what social groups robots represent) or ethical issues of relationship (human-like form in robots enables to replace relationships with human beings).

This book focuses on two categories of ethical issues: those that are modified by human likeness and those that are unique to human-like robots. Due to the short format of this book, I focus on selected issues. Some of the issues covered are centrally exposed as main themes of chapters, such as moral patience or representational issues. However, others are somewhere in between, like deception or ethical issues around sex robots. It needs to be underlined that this book does not cover all ethical issues that might be related to human-like robots. The selection of the topics was also dictated by my previous research interests. The overview of the issues covered in this book is given in the section "Structure and Approach."

The Ethical Meaning of the Design of Human-like Robots

The third group of matters that needs to be unpacked here relates to the word "design." This word is used here in two main, interrelated senses. First, to underline that we are not yet there, and there is still time to introduce some changes; second, to refer to approaches that are focused on embedding values in the design of technologies, including robots. I will now explain these two senses in detail.

The ethical deliberations focus on the variant life cycles of technologies and their deployment and use, in addition to design. By highlighting "design," I want to pay attention to the time frame in which we are now regarding human-like robots. We are at the beginning of their deployment. We are still in the position that some negative consequences might be stopped or reduced. By talking about design in this context, we also indirectly pay attention to the designer. There is someone who has the power to shape robots in one or another way. The process of deploying robots is not happening impersonally, as such. Other actors are involved in the whole process, who need to recognize their agency in the process of deploying human-like robots. For example, Zawieska encourages roboticists who directly shape robots to engage with ethical issues, pointing out that "the risk of dehumanization is inherent to the research that aims to recreate humans in artificial systems" (Zawieska 2020: 878).

The second sense in which design is used here is to acknowledge approaches that focus on embedding human values into the design process like "ethics by design" (see, e.g., Dodig Crnkovic and Çürüklü 2012; Dodig-Crnkovic et al. 2021), "values-sensitive design" (see, e.g., Friedman and Hendry 2019; Cheon and Su 2016; Umbrello et al. 2021), or "integrative social robotics." The last approach underlines the need to implement robots that are culturally sustainable with awareness of the cultural dimensions (see, e.g., Seibt et al. 2018, 2020). To put it simply, the approaches mentioned emphasize that technologies in general, and robots in particular, should be created with human values in mind during the planning phase. This book, by focusing on various ethical risks, emphasizes the issues that should be discussed while designing robots.

Structure and Approach

This book has six chapters. The first chapter serves as an introduction which explains the motives for writing this book and unpacks the key concepts used within. Chapters 2 to 5 cover specific ethical issues related to human likeness of robots. These chapters can be read independently to some extent. In those chapters, I start with the scenarios that aim to illustrate the matter under discussion. The final chapter presents the conclusions. The beginning of every chapter has an illustration that gives an idea of its content. It is followed by a short overview of the ideas discussed in the chapter.

The second chapter discusses the impact of the human-like form of robots on the safety of humans. The source of problems is not the behavior of robots, but their presence among human beings. So far, the only human-shaped entities around are humans themselves. We establish rules covering how we should treat each other to ensure mutual safety; the presence of another entity that looks like us has a disruptive potential. The chapter focuses on two risk categories that are related to the human form: direct and indirect. Direct risks concern the impact of the presence of human-like robots in humans' immediate surroundings. We could mistake robots for humans, for example, in traffic when there is a need to make a decision regarding a crash (an epistemological threat). We might also be aware that human-like robots are just robots, but we might develop an attachment to them, which could lead to extreme circumstances of harm to humans by prioritizing robots instead of humans (a patient threat). The wider deployment of human-like robots in common spaces could also lead to indirect risks. One such risk is related to changing the standards of interactions among humans for the worse due to interactions with robots. One of the topics that is discussed in this context is related to sex robots and the potential impact due to the transferring of their experience to humans. Another indirect risk is related to the devaluation of human shape in public spaces. The presence of a human-shaped object that is not moving in the case of humans should trigger concerns about potential health problems and initiate a check up to see whether the person needs professional help. But the common presence of human-like robots might decrease the responses that might save humans in danger.

The third chapter is concerned with the problem of the representation of human-like robots. Choosing human form entails the possibility of social categorization of robots. Human-like robots could look not just like humans but like humans of a specific age, gender, and race, to name a few categorizations. The choice of a specific categorization might be useful from the perspective of the performance of robots. However, there are also risks related to the harm that could be done to social groups of humans by representing them in robots. Studies suggest that humans adopt human stereotypes to robots. It is pointed out that interactions with robots might strengthen existing stereotypes, which could lead to harm to humans. This effect should not be ignored; robots might impact equality and diversity within society, and because of that it is an issue that should be within the orbit of political bodies' interests to ensure that the deployment of robots will be beneficial. The chapter also discusses serious strategies for responding to the risk related to representational harms, including calls to abandon human form from robots that aim to do tasks that could be seen as degrading.

The fourth chapter changes the matters of concern from issues related to external resemblance, that which is easily observable, to a discussion of attributes that are more related to the internal aspects of human likeness that are harder to grasp. Chapter 4 deals with the limits of making robots human like. Specifically, there are two questions. First, could robots have human-like ethics; second, could robots be responsible in a human-like sense? The chapter problematizes the answer to these two questions. In the case of human-like ethics, it is pointed out that ethics is a complex endeavor that might be limited in case of human-like entities, which means that robots cannot be fully ethical in the human sense. One of the main obstacles is the lack of emotions in robots. The project of making robots ethical needs to be inevitably less ambitious. Such a conclusion entails that having robots able to make decisions marked by ethics puts burdens on humanity, which would always need to supervise and change the guiding ethical principles to make robots adapt to changing moral realities. The answer to the question of responsibility is also problematic. The dominant view is that robots cannot be responsible in a human-like sense yet, and some believe that it will never be possible. That strong view is related to the notion of moral responsibility in the classical sense, which requires moral agents to stand behind the action for which responsibility is to be ascribed. If robots do not have the qualities that we associate with entities that are responsible, then their responsibility is excluded. However, there are positions according to which robots might be responsible, at some point. Even if the latter is true, the responsibility would not be identical to the responsibility that humans have, with the exact same punishments as humans are exposed to. There are also positions that discuss issues of responsibility that focus on responsibility practices and not that much on the qualities of the agent. Having such a starting point, to some extent the robots could be included in the responsibility practices.

The fifth chapter talks about how humans should treat human-like robots. The answer to this problem is divided into two options. First, robots are human-like in the sense that they share some internal human-like properties with humans. Second, they share only external shells with humans. In the case of robots with human-like qualities, the chapter considers various ethics-related issues. There are problems with what should count as qualities that make robots human like, such as intelligence, consciousness, or something else. There is also the problem of how we know that robots have such qualities. There are issues with the ethics of reasons for making robots that are human like. Should we bring them into a world of suffering? Should we have other entities with whom we would need to share resources and space? Should we force robots to live their lives to serve humans? Having human-like robots might entail several burdens on both individuals and societies, and be intrinsically unethical. Ethical issues also arise when robots' human likeness is limited to their appearance. There are other reasons why we should care about such robots. It might be necessary to treat human-like robots with caution to prevent harm to humans, and there could be problems distinguishing robots from humans. There are also formulated concerns that the mistreatment of robots could transfer to the mistreatment of human beings. The other issue is that an external observer could have unpleasant feelings while witnessing abuse of human-like robots; they could not know that robots are under attack, even if the attacker is fully aware of this fact. Human-like robots might also be a platform for the development of certain relationships that are otherwise possible to develop between humans, and which society considers to be important. In this context, such relationships as friendship and love are often discussed. If humans could develop such a relationship with robots, then there may be a need to protect those relations by protecting the robots with which such relations have been developed.

The sixth chapter concludes this book. It brings together the discussions in the other chapters and arranges the specific ethical issues of human likeness into three categories of more general inputs. While the book conveys the message that the human-like robot project is inherently disruptive, at the same time it also acknowledges that stopping the whole project is not possible and also not desirable. Despite their drawbacks, there are applications for humanoid robots that might be useful. First, out of the three categories mentioned, there are aspects of human likeness and types of human-like robots that might be considered banned. Second, some specific aspects of human likeness should be used with caution. There are also positive interventions that should accompany the deployment of human-like robots. Changes should happen at the individual level, in the design of human-like robots, as well as in the moral and legal systems of societies. The sixth chapter also briefly discusses the ways in which ethical deliberations could be transferred into the practice of human-like robots. It highlights the multidimensionality of means by which the regulation should happen, as well as the multitude of actors that should be engaged. The chapter also places attention on the role of the people engaged in the development of robots and calls for roboactivism.

This book draws on literature from several fields such as human–robot interactions, ethics of artificial intelligence (AI) and robotics, and the philosophy of technology and law. Due to numerous publications from all these fields dealing with issues of human-like robots, and due to the limited length of this book, I had to narrow down the topics covered, as already mentioned, and also the sources that I cite throughout the book. The reader needs to be aware that there is much more relevant literature covering the issues discussed here. For example, there is growing literature containing non-Western ethical and philosophical perspectives on technology in general, and robotics in particular, which I have barely mentioned here and which deserves more attention (see, e.g., Zhu et al. 2019; Gellers 2020; Jecker 2021; Wong and Wang 2021; Alemi et al. 2020; Hongladarom and Bandasak 2023; Hongladarom 2020; Friedman 2022).

With this book, I have continued my interest in the impact of technology, in particular of robots, on humans and society. My last book was also about robots (Mamak 2023) wherein I showed that the question "What is wrong with kicking a robot?" is relevant from the perspective of criminal law. Criminal law could give multiple answers to this question. Robots embed values protected by criminal law, and attacks on robots that hold values could be an assault on a range of legally protected values that are already safeguarded or should be safeguarded. While conducting research for my previous book, I developed initial ideas that were beyond the scope of that project; I have finally developed them and found a place for them in the present book. I also draw upon on my other works: journal articles, chapters, collected volumes, and conference proceedings; each time when I build upon them, I cite the relevant publications.

References

Alemi, Minoo, Alireza Taheri, Azadeh Shariati, and Ali Meghdari. 2020. "Social Robotics, Education, and Religion in the Islamic World: An Iranian Perspective." *Science and Engineering Ethics* 26 (5): 2709–34. https://doi.org/10.1007/s11948-020-00225-1.

Alsegier, Riyadh A. 2016. "Roboethics: Sharing Our World with Humanlike Robots." *IEEE Potentials* 35 (1): 24–28. https://doi.org/10.1109/MPOT.2014.2364491.

Asaro, Peter M. 2006. "What Should We Want from a Robot Ethic?" *International Review of Information Ethics* 6 (12): 9–16.

Bartneck, Christoph, Christoph Lütge, Alan Wagner, and Sean Welsh. 2021. *An Introduction to Ethics in Robotics and AI*. SpringerBriefs in Ethics. Cham: Springer International Publishing. https://doi.org/10.1007/978-3-030-51110-4.

Brey, Philip A. E. 2012. "Anticipatory Ethics for Emerging Technologies." *NanoEthics* 6 (1): 1–13. https://doi.org/10.1007/s11569-012-0141-7.

Calo, Ryan. 2012. "Robots and Privacy." In *Robot Ethics: The Ethical and Social Implications of Robotics*, edited by Patrick Lin, Keith Abney, and George A. Bekey, 187–201. MIT Press. https://ieeexplore.ieee.org/document/6733953.

Čapek, Karel. 2015. *R.U.R. Rossum's Universal Robots*. Marsilio.

Cheon, EunJeong, and Norman Makoto Su. 2016. "Integrating Roboticist Values into a Value Sensitive Design Framework for Humanoid Robots." In *2016 11th ACM/IEEE International Conference on Human–Robot Interaction (HRI)*, 375–82. https://doi.org/10.1109/HRI.2016.7451775.

Coeckelbergh, Mark. 2020. *AI Ethics*. Cambridge, MA: MIT Press. https://mitpress.mit.edu/books/ai-ethics.

———. 2022. *Robot Ethics*. Cambridge, MA: MIT Press.

Collingridge, David. 1981. *Social Control of Technology*. New edition. Milton Keynes: Open University Press.

Darling, Kate. 2021. *The New Breed: How to Think about Robots*. Dublin: Allen Lane. https://us.macmillan.com/thenewbreed/katedarling/9781250296115.

Devlin, Kate. 2018. *Turned On: Science, Sex and Robots*. Illustrated edition. London: Bloomsbury Sigma.

Dodig Crnkovic, Gordana, and Baran Çürüklü. 2012. "Robots: Ethical by Design." *Ethics and Information Technology* 14 (1): 61–71. https://doi.org/10.1007/s10676-011-9278-2.

Dodig-Crnkovic, Gordana, Tobias Holstein, and Patrizio Pelliccione. 2021. "Future Intelligent Autonomous Robots, Ethical by Design. Learning from Autonomous Cars Ethics." arXiv. https://doi.org/10.48550/arXiv.2107.08122.

Estrada, Daniel. 2020. "Human Supremacy as Posthuman Risk." *The Journal of Sociotechnical Critique* 1 (1): 1–40. https://doi.org/10.25779/j5ps-dy87.

Fink, Julia. 2012. "Anthropomorphism and Human Likeness in the Design of Robots and Human–Robot Interaction." In *Social Robotics*, edited by

Shuzhi Sam Ge, Oussama Khatib, John-John Cabibihan, Reid Simmons, and Mary-Anne Williams, 199–208. Lecture Notes in Computer Science. Berlin, Heidelberg: Springer. https://doi.org/10.1007/978-3-642-34103-8_20.

Fitzpatrick, Paul, Kensuke Harada, Charles C. Kemp, Yoshio Matsumoto, Kazuhito Yokoi, and Eiichi Yoshida. 2016. "Humanoids." In *Springer Handbook of Robotics*, edited by Bruno Siciliano and Oussama Khatib, 1789–1818. Springer Handbooks. Cham: Springer International Publishing. https://doi.org/10.1007/978-3-319-32552-1_67.

Friedman, Batya, and David Hendry. 2019. *Value Sensitive Design: Shaping Technology with Moral Imagination*. Cambridge, MA: MIT Press.

Friedman, Cindy. 2022. "Ethical Concerns with Replacing Human Relations with Humanoid Robots: An Ubuntu Perspective." *AI and Ethics* 3: 527–38. https://doi.org/10.1007/s43681-022-00186-0.

Gasser, Georg. 2021. "The Dawn of Social Robots: Anthropological and Ethical Issues." *Minds and Machines* 31 (3): 329–36. https://doi.org/10.1007/s11023-021-09572-9.

Gellers, Joshua C. 2020. *Rights for Robots: Artificial Intelligence, Animal and Environmental Law*. London: Routledge. https://doi.org/10.4324/9780429288159.

Gensler, Harry J. 2011. *Ethics: A Contemporary Introduction*. 2nd edition. New York: Routledge. https://doi.org/10.4324/9781351231831.

Gunkel, David J. 2018. *Robot Rights*. Cambridge, MA: MIT Press.

———. 2023. *Person, Thing, Robot: A Moral and Legal Ontology for the 21st Century and Beyond*. Cambridge, MA: MIT Press.

Guthrie, Stewart Elliott. 1997. "Anthropomorphism: A Definition and a Theory." In *Anthropomorphism, Anecdotes, and Animals*, edited by Robert W. Mitchell, Nicholas S. Thompson, and H. Lyn Miles, 50–58. SUNY Series in Philosophy and Biology. Albany, NY: State University of New York Press.

Hakli, Raul, Pekka Mäkelä, and Johanna Seibt, eds. 2023. *Social Robots in Social Institutions: Proceedings of Robophilosophy 2022*. IOS Press.

Heater, Brian. 2024. "Sanctuary's New Humanoid Robot Learns Faster and Costs Less." *TechCrunch* (blog), April 25. https://techcrunch.com/2024/04/25/sanctuarys-new-humanoid-robot-learns-faster-and-costs-less/.

Hongladarom, Soraj. 2020. *The Ethics of AI and Robotics: A Buddhist Viewpoint*. Rowman & Littlefield.

Hongladarom, Soraj, and Jerd Bandasak. 2023. "Non-Western AI Ethics Guidelines: Implications for Intercultural Ethics of Technology." *AI & SOCIETY*, April. https://doi.org/10.1007/s00146-023-01665-6.

Howe, Ben Ryder, and Nic Antaya. 2023. "The Robots We Were Afraid of Are Already Here." *The New York Times*, July 29, sec. Business. https://www.nytimes.com/2023/07/29/business/robots-workers-future.html.

Huber, Nick. 2024. "'Humanoid' Robot Wave Signals Change on the Production Line." *Financial Times*, March 26, sec. AI and Jobs. https://www.ft.com/content/613cb2c6-7067-413f-a3cf-747519b407c2.

Jecker, Nancy S. 2021. "Can We Wrong a Robot?" *AI & SOCIETY*, November. https://doi.org/10.1007/s00146-021-01278-x.

Jordan, John M. 2016. *Robots*. 1st edition. Cambridge, MA: MIT Press.

Kolodny, Lora. 2023. "Agility Robotics Is Opening a Humanoid Robot Factory, Beating Tesla to the Punch." CNBC, September 18. https://www.cnbc.com/2023/09/18/agility-robotics-is-opening-a-humanoid-robot-factory-.html.

Kudina, Olya, and Peter-Paul Verbeek. 2019. "Ethics from Within: Google Glass, the Collingridge Dilemma, and the Mediated Value of Privacy." *Science, Technology, & Human Values* 44 (2): 291–314. https://doi.org/10.1177/0162243918793711.

Łichocki, Paweł, Aude Billard, and Peter H. Kahn. 2011. "The Ethical Landscape of Robotics." *IEEE Robotics & Automation Magazine* 18 (1): 39–50. https://doi.org/10.1109/MRA.2011.940275.

Lin, Patrick. 2012. "Introduction to Robot Ethics." In *Robot Ethics: The Ethical and Social Implications of Robotics*, edited by Patrick Lin, Keith Abney, and George A. Bekey, 3–15. Cambridge, MA: MIT Press. https://ieeexplore.ieee.org/document/6734015.

Mamak, Kamil. 2023. *Robotics, AI and Criminal Law: Crimes against Robots*. Routledge.

Mann, Jyoti. 2023. "China Boldly Claims It Has a Plan to Mass-Produce Humanoid Robots That Can 'Reshape the World' within 2 Years." *Business Insider*, 7 November. Accessed May 1, 2024. https://www.businessinsider.com/china-plans-mass-production-humanoid-robots-within-two-years-2023-11.

Mara, Martina, Markus Appel, and Timo Gnambs. 2022. "Human-Like Robots and the Uncanny Valley." *Zeitschrift Für Psychologie* 230 (1): 33–46. https://doi.org/10.1027/2151-2604/a000486.

Mayor, Adrienne. 2018. *Gods and Robots: Myths, Machines, and Ancient Dreams of Technology*. Princeton University Press. https://press.princeton.edu/books/hardcover/9780691183510/gods-and-robots.

Mori, M. 1970. "Bukimi No Tani [the Uncanny Valley]." *Energy* 7: 33–35.

Nyholm, Sven. 2020. *Humans and Robots: Ethics, Agency, and Anthropomorphism*. Illustrated edition. London, New York: Rowman & Littlefield Publishers.

de Pagter, Jesse. 2023. "Ethics and Robot Democratization: Reflecting on Integrative Ethics Practices." *International Journal of Social Robotics* 15 (12): 2005–18. https://doi.org/10.1007/s12369-023-01005-0.

Perugia, Giulia, Stefano Guidi, Margherita Bicchi, and Oronzo Parlangeli. 2022. "The Shape of Our Bias: Perceived Age and Gender in the Humanoid Robots of the ABOT Database." In *Proceedings of the 2022 ACM/IEEE International Conference on Human–Robot Interaction*, 110–19. HRI '22. Sapporo, Hokkaido: IEEE Press.

Putman, Hilary. 1964. "Robots: Machines or Artificially Created Life?" *Journal of Philosophy* 61 (21): 668–91. https://doi.org/10.2307/2023045.

Roth, Emma. 2024. "Boston Dynamics Atlas Robot: A Full History." *The Verge*, April 16. https://www.theverge.com/24132451/boston-dynamics-atlas-robot-retirement.

Russell, Stuart. 2019. *Human Compatible: AI and the Problem of Control*. Allen Lane.

Seibt, Johanna, Malene Flensborg Damholdt, and Christina Vestergaard. 2018. "Five Principles of Integrative Social Robotics." *Frontiers in Artificial Intelligence and Applications* 311: 28–42. https://doi.org/10.3233/978-1-61499-931-7-28.

———. 2020. "Integrative Social Robotics, Value-Driven Design, and Transdisciplinarity." *Interaction Studies. Social Behaviour and Communication in Biological and Artificial Systems* 21 (1): 111–44. https://doi.org/10.1075/is.18061.sei.

Shafer-Landau, Russ. 2020. *The Fundamentals of Ethics*. 5th edition. New York: Oxford University Press.

Spatola, Nicolas, Sophie Monceau, and Ludovic Ferrand. 2020. "Cognitive Impact of Social Robots: How Anthropomorphism Boosts Performances." *IEEE Robotics & Automation Magazine* 27 (3): 73–83. https://doi.org/10.1109/MRA.2019.2928823.

Sriram, Akash. 2024. "Tesla Could Start Selling Optimus Robots by the End of Next Year, Musk Says." *Reuters*, April 24, sec. Technology. https://www.reuters.com/technology/tesla-could-start-selling-optimus-robots-by-end-next-year-musk-says-2024-04-24/.

Sweeney, Paula. 2023. *Social Robots: A Fictional Dualism Model*. Rowman & Littlefield Publishers, Incorporated.

Turner, Jacob. 2019. *Robot Rules: Regulating Artificial Intelligence*. Cham: Springer International Publishing. https://doi.org/10.1007/978-3-319-96235-1.

Tzafestas, Spyros G. 2015. *Roboethics: A Navigating Overview*. Springer.

Umbrello, Steven, Marianna Capasso, Maurizio Balistreri, Alberto Pirni, and Federica Merenda. 2021. "Value Sensitive Design to Achieve the UN SDGs with AI: A Case of Elderly Care Robots." *Minds and Machines* 31 (3): 395–419. https://doi.org/10.1007/s11023-021-09561-y.

Veruggio, Gianmarco. 2005. "The Birth of Roboethics." In *ICRA 2005, IEEE International Conference on Robotics and Automation, Workshop on Roboethics*, edited by John Doe, 1–4. https://philarchive.org/rec/VERTBO-3.

Veruggio, Gianmarco, Fiorella Operto, and George Bekey. 2016. "Roboethics: Social and Ethical Implications." In *Springer Handbook of Robotics*, edited by Bruno Siciliano and Oussama Khatib, 2135–60. Springer Handbooks. Cham: Springer International Publishing. https://doi.org/10.1007/978-3-319-32552-1_80.

Weng, Yueh Hsuan, and Yasuhisa Hirata. 2018. "Ethically Aligned Design for Assistive Robotics: 2018 International Conference on Intelligence and Safety for Robotics, ISR 2018." *2018 International Conference on Intelligence and Safety for Robotics, ISR 2018*, November, 286–90. https://doi.org/10.1109/IISR.2018.8535889.

Winfield, Alan. 2012. *Robotics: A Very Short Introduction*. Very Short Introductions. Oxford, New York: Oxford University Press.

Wong, Pak-Hang, and Tom Xiaowei Wang. 2021. *Harmonious Technology: A Confucian Ethics of Technology*. Routledge.

Wynsberghe, Aimee van. 2016. *Healthcare Robots: Ethics, Design and Implementation*. Abingdon, Oxon: Routledge.

Zawieska, Karolina. 2020. "Disengagement with Ethics in Robotics as a Tacit Form of Dehumanisation." *AI & SOCIETY* 35 (4): 869–83. https://doi.org/10.1007/s00146-020-01000-3.

Zhu, Qin, Tom Williams, and Ruchen Wen. 2019. "Confucian Robot Ethics." *Computer Ethics—Philosophical Enquiry (CEPE) Proceedings* 2019 (1). https://doi.org/10.25884/5qbh-m581.

Zlotowski, J., D. Proudfoot, and Christoph Bartneck. 2013. "More Human Than Human: Does the Uncanny Curve Really Matter?" http://hdl.handle.net/10092/8698.

Złotowski, Jakub, Diane Proudfoot, Kumar Yogeeswaran, and Christoph Bartneck. 2015. "Anthropomorphism: Opportunities and Challenges in Human–Robot Interaction." *International Journal of Social Robotics* 7 (3): 347–60. https://doi.org/10.1007/s12369-014-0267-6.

2 | Risks of Living with Human-like Robots

Introduction

The fulfillment of the dream of creating human-like entities is closer than ever, thanks to the development of robots. Giving robots a human form might be useful at many levels, but it also involves risks. To illustrate the complexity of the risks let us start with a few scenarios.

You are driving in your car, thinking about the dinner that you will be eating soon, but suddenly, your good mood disappears—you realize that the brakes are malfunctioning. You desperately try to activate them, but nothing works. You know that you need to crash your car to stop it. You look nervously at the road and its surroundings and see that there is no way to avoid hitting humans. On one side, there are two people going, and on the other there is only one person. You have only a second to decide, and you turn your car in the direction of the single person, which seems to be the most rational choice in that tragic situation. Unfortunately, the person that you hit with the car did not survive. You are okay. The accident is being investigated, and after carefully reviewing the circumstances from various cameras, you are informed that the two people on the other side of the road were humanoid robots. Robots are quite easy to distinguish from humans in normal settings, but in the given situation that precedes an accident, in limited daylight, rain, and in a rush, it is understandable that they could be mistaken for humans.

In the second scenario, you are hiking with a group of people in the mountains. The weather is initially pretty good, but it suddenly changes, as can happen in the mountain areas. There are heavy rains and strong winds. Someone from the group notices that rocks that broke away from the upper part of the mountain are coming in your direction at a high speed. You are happy to find shelter quickly in which there is a space for everyone from the group. Unfortunately, you notice that one person is still in danger, trying to carry the humanoid robot that stopped working. That person left behind their backpack and other belongings but could not leave the robot behind. It was too late to seek shelter when that person decided to give up the robot. If the person had chosen to leave the robot earlier, they would have survived.

In the third scenario, a child spent a couple of his early years interacting with a humanoid robot. He was an only child, and his parents wanted him to have a companion to play with, so they decided to buy a robot. The robot looked like a human child. When the family moved to another city, the child went to preschool for the first time in his life. At the end of the first day, the parents were informed that their son behaved violently toward other children, hitting them. The parents were worried and decided to install cameras to observe the way in which their child interacted with the robot. They found out that the child was not aggressive, but in physical interactions he needed to use more muscle force to play than would be acceptable with a human child. Playing in the exact same way with robots was okay but replicating that behavior with other children was considered by them to be violence.

In the fourth scenario, you are going home after a long day at work. With a lot of ideas, you approach your car, which is parked a few minutes away. You are listening through headphones to your favorite songs that help you to relax. Lying under the bench in the park next to the parking lot, you notice a humanoid robot; your first thought is, "Another robot this week. They are making worse and worse batteries these days," and you go back to listening to your song. The next day, when you are back in the same parking place, you notice an ambulance and a police car near the bench. With curiosity, you approach the place and ask questions about what happened. You are told that this was a man who was in pain the whole night; he finally died early in the morning. You realize that it was the person that you had mistaken for a robot yesterday, and what is worse, if you had called an ambulance yesterday, then there was a chance that he would have survived.

All these scenarios point to the fact that the human likeness of robots contributes to the harm that is done to human beings. Robots did not cause that harm personally, and the intention of the people who acted the way they did was not necessarily wrong: the car driver wanted to avoid increased damage in a crash situation; the person who died on the mountain wanted to save their beloved companion; a preschool child recreated the patterns of interactions that he learned by playing with a humanoid robot; and the person who left a dying human thinking that it was a robot in need of repair. Nothing required their intervention, like noticing the electric scooter left in a place that it was not supposed to be in. The robots themselves did not cause harm, but their presence in humanoid form in the social circle made it possible for those things to happen. This chapter explores the risks that could follow by choosing the human form in robot design.

It seems uncontroversial to say that all robots should be safe. It does not matter whether it is a vacuum cleaner, an autonomous car, or an industrial robot. Like other products, before being deployed in the market, robots should meet certain criteria, be made from appropriate materials, be tested, obtain certificates, and so on. (On robots as products, see, e.g., Bertolini 2013.) Humanoid robots, due to their potential application in almost every sphere of our lives, could embody and also increase all safety problems that are known from other kinds of robots (Veruggio et al. 2016: 2149). Safety standards might also vary between domains. For example, robots in hospitals might be required to be made of safer components than robots deployed in industrial settings (specifically, on healthcare robots, see, e.g., Wynsberghe 2016; Fosch-Villaronga 2019). The interactions with people with reduced immunity, among viruses and infectious diseases, make healthcare robots riskier compared with robots that are deployed in factories with limited interactions with humans. However, despite the differences, some level of safety is expected from all robots.

One of the typical ways of thinking about the safety of robots is to be concerned about how they operate. In other words, we worry whether robots will act in a manner that will not put us in danger. In ethical literature, it is usually related to the topic of machine/robot ethics (see, e.g., Wallach and Allen 2009). It will be a matter of broader discussion in Chapter 4. In this chapter, my focus is on human likeness as a unique problem for human safety. The sole presence of human-like entities among us is socially disruptive.

Disruptiveness in technologies has been used for some time to underline their disturbing impact on the domains in which they are deployed and on society at large. As van de Poel et al. in their book about the ethics of socially disruptive technologies define, "social disruptions may be understood as changes that prevent important aspects of human society (broadly understood) from continuing without change, thereby generating disorder or upheaval" (van de Poel et al. 2023: 16). The population of the world with entities that look like us but are not like us could change the way in which we interact. As a society, we should respond to that, or at least be aware of this aspect of the decision to make robots in human form.

Before going further, some explanatory remarks need to be made. Human likeness in the context of this chapter is the appearance of the robots, not their internal qualities. The discussion could be different if robots had other properties that we thought were important from a moral point of view (Mamak 2021). However, in the contexts presented in this chapter, I assume that human-like robots share only the external shell with humans, and not their internal aspects. The meaning of the internal qualities of robots for the purpose of their treatment is discussed in Chapter 5. Presenting the issues with human-likeness in this way makes the issues discussed here more urgent, compared to the discussion on what would happen if humans like to share qualities similar to humans. It means that some of the problems described here can happen even now, with the current state of technological developments. There are already robots that more or less look like humans.

The aforementioned scenarios introduced the types of issues that will be discussed here. In the following, there is a more systematic description of the issues. The risks that are discussed in this chapter are divided into two categories: direct and indirect. These categories will be examined in the two sections that follow. The last part of this chapter presents the conclusion.

Direct Risks

By direct risks, I mean that humans are endangered by the presence of human-like robots. In other words, humans might be hurt because they are unlucky to be in the same place that humanoid robots are. Elsewhere, I categorized them into two types of threats: "epistemological threat" and "patient threat" (Mamak 2024; see also Mamak 2021). I formulated them based on the observation that anthropomorphic robots could change the way in which the value of human life is protected by the law. It is pointed out that one of the basic aims of the state is to protect the members of the community (see, e.g., Heyman 1994; Ashworth 1975; Kant 1991: 136), and from that perspective, the emergence of human-like nonhumans is problematic because it disturbs the way things are going in that regard.

Epistemological Threat

The epistemological threat relates to our limited ability to collect information about the external world. To put it in simple words, we could mistake robots for humans, and because of that mistake, humans might be harmed. Our biological apparatus allows us to sense words only to some extent. The main element of the sense apparatus is vision. According to Ripley and Politzer, "[i]t is estimated that up to eighty percent of our perception, learning, cognition and activities are mediated at least to some extent through vision" (Ripley and Politzer 2010: 215). In some contexts, we can only relate to that sense. In the first scenario mentioned at the beginning of this chapter, the driver hit a human being, saving two humanoid robots that were on the other side of the road; the driver thought that those robots were humans. The driver could not touch, smell, or talk to the robots; the decision was solely based on the material delivered optically. The epistemological threat could materialize in the case of robots that differ drastically in human likeness. In a context in which it is needed to recognize humans from a distance, like in the aforementioned scenario on the road, human likeness could be less sophisticated to confuse a person who needs to make morally important decisions. Humanoid robots might also be almost indistinguishable from humans on another extremum of the scale of human likeness. The more human-like the robot is, the harder it is to distinguish it from humans visually, making it difficult to decide who will be the injured party—robot or human.

If we struggle to say whether the figure in front of us is human or humanoid robot, then it is close to the idea that is popular in philosophical literature about philosophical zombies (Kirk 1974). According to Chalmers, the zombie is "someone or something physically identical to me (or to any other conscious being), but lacking conscious experiences altogether" (Chalmers 1996: 94). The idea of a zombie is used in philosophy to help illustrate various philosophical problems (see, e.g., Kirk 2021). It is also used as an idea in the context of robots (see, e.g., Hansen 2023). One thing that might be asked when we do not know whether the entity that we are approaching is a human or humanoid robot is how to treat such an entity (Danaher 2020). The uncertainty about the humanness of such entities is also important from the perspective of making preferences in favor of human beings. In making moral decisions, human beings should be preferred, and having robots with human likeness hampers the process of making just decisions, even if we want to prefer humans over robots.

Another aspect of the epistemological threat is when one expects to interact with humanoid robots, but instead humans appear. There could be contexts in which mistreating human-like robots might be acceptable, and then the presence of human beings in such places might be dangerous for them. For example, let us imagine that there is a place where shooting skills are being practiced on human-like robots. I do not claim here that such behaviors are morally unproblematic, but it could be imagined that such a situation might take place. What if a human-like object appears in a place in which there is shooting, where it is reasonable to expect that it is a humanoid robot and not a human? What if there is a hunt in the forest for boars and if a human being wears a realistic boar costume, is on their hands and knees, and starts to make boar-like noises, would it be justifiable to mistake such a person for a boar? Pretending to be a boar, however, would require effort. If it is permissible to shoot at humanoid robots, then the person accidentally mistaken for the object of the hunt does not need to do anything to be mistaken as a robot; they just need to be in the wrong place at the wrong time.

Patient Threat

The other direct risk related to the human likeness of robots is patient threat. This threat materializes when humans, who are going to make morally important decisions while considering ethical variables, bring robots into the equation, which might put humans in danger. The second scenario from the beginning of this chapter is related to this threat. In that scenario, a person who stops to help a robot, instead of taking shelter in the mountains in a deadly situation, treats the robot not as a mere thing but as something more. This threat differs from the previous one—epistemological. The decision-maker knows that they are dealing with robots, but despite that knowledge, what they decide could harm human beings, not only themselves, as it was in that scenario, but also others.

When robots are treated other than as mere tools, it might be a sign that they are considered to be moral patients. Moral patients will be discussed more in Chapter 5. For now, it should be clarified that the moral patient is an entity that belongs to a moral circle; it is a receiver of morally relevant actions, an entity that might be morally harmed. As Levy and Savulescu put it, "[t]o be a moral patient is to be a being whose welfare matters, whose welfare must be taken into account when we decide what to do" (Levy and Savulescu 2009, 366). In the context of robots, if we consider those entities to be moral patients, then the acceptable range of behaviors toward them might be limited compared to behaviors performed on tables and chairs.

Treating humanoid robots in a kind way is a positive thing to do and may be called virtuous (but see Bryson 2018). On the other hand, wronging robots might be morally problematic (for an overview, see Gunkel 2018, 2023), and could even deserve criminal treatment (Mamak 2023). However, when human welfare is at stake due to the treatment of robots as moral patients, then it is problematic and deserves attention.

Various studies have suggested that humans treat robots not as mere things, but as something more. Prescott expressed that kind of thinking about machines in his paper "Robots are not just tools," in which he considers the phrase "robots are just tools" as inappropriate considering human responses to robots (Prescott 2017). The body of research on human–robot interaction sheds light on how humans respond to robots. In relatively early research on mistreatment in the context of human–robot interaction, Salvini and colleagues suggested that the language of abuse better describes the nature of such behaviors rather than vandalism (Salvini et al. 2010). A study authored by Carlson et al. showed that there was a difference in perception of mistreatment between robots and computers; participants felt more sympathetic toward robots than computers (Carlson et al. 2019). In a study conducted by Bartneck and Keijsers, participants were exposed to videos showing abusive behavior toward humans and robots. The task was to rate the moral acceptability of the actions. Participants expressed no significant difference in the expression of the immorality of those behaviors toward human and robot victims (Bartneck and Keijsers 2020). Various studies have shown that humans feel empathy to robots (see, e.g., Suzuki et al. 2015; Rosenthal-von der Pütten et al. 2013, 2014; on empathy toward robots, see also Malinowska 2021a, 2021b). However, it does not mean that people respond in exactly the same way to the abuse of robots as they do to similar behaviors performed against humans. For example, a study conducted by Sanoubari et al. shows that while mistreatment of both robots and humans was described by participants as morally wrong, there were differences in responses; for example, participants were less willing to intervene when the robot was mistreated (Sanoubari et al. 2021). In a review paper on prosocial human responses to robots, Nielsen et al. noted that humans do not treat robots as humans but as "somewhat human," not in binary categories—human or not human—but as more or less human (Nielsen et al. 2022).

It is crucial to analyze studies that focus on the manipulation of the level of human likeness and its impact on human responses to robots and people. Krach et al. (2008) investigated how human likeness affects the ability to attribute intentions and desires (Theory of Mind) to machines, specifically whether increasing the human likeness of interaction partners modulates participants' Theory of Mind–associated cortical activity. The findings revealed that the tendency to build a model of another's mind grows linearly with perceived human likeness (Krach et al. 2008). The more a robot is human-like, the more we attribute human qualities to it. (For a review of the impact of anthropomorphism, see, e.g., Spaccatini et al. 2023: 2–3.) According to a study by Riek et al., people empathize more strongly with more human-looking robots and less with mechanical-looking robots (Riek et al. 2009). However, it should be noted that the level of human likeness in robots could have a different impact based on how people respond to robots in different contexts. For example, in a study conducted by Złotowski et al., "a highly humanlike robot is perceived [as being] less trustworthy and empathic than a machine-like robots with some humanlike features" (Złotowski et al. 2016: 64). Nevertheless, when human safety is at stake, increased human likeness could be considered to be for a danger to human safety, which will now be explained.

There are studies that have shown the potential risks to human life for being "too nice" to human-like robots. In a study conducted by Nijssen with colleagues, the researchers wanted to examine the impact of anthropomorphic appearance on people's utilitarian decision-making about robotic agents. They used dilemmas that were structured according to this logic:

"A group of people is in danger of dying or getting seriously injured, but they can be saved if the participant decides to perform an action that would mean sacrificing an individual agent (human, human-like robot, or machine-like robot) who would otherwise remain unharmed" (Nijssen et al. 2019: 45–46).

According to the results of this experiment, some people hesitated to sacrifice robots to save humans. What is important in the context of this book, responses varied with perceptual humanness: "agents who looked least human (machine-like robots) were found to be sacrificed more often in moral dilemmas than humans" (Nijssen et al. 2019: 47) and "Robots that were perceptually more human-like were sacrificed less often compared to their less human-like counterparts. [...] participants in our study decided to rather save a humanized robot instead of saving a group of anonymous people by killing that robot, which is a remarkable and morally relevant finding" (Nijssen et al. 2019: 51–52). The authors noticed that the way in which humans respond to robots is morally wrong, which might affect human beings. In another study led by Nijssen, the authors wanted to examine whether a similar effect would be noticeable in less dramatic everyday situations. They found that the effect of anthropomorphization in such contexts is different, "anthropomorphizing a robot may lead us to save it when it is about to perish, it does not make us more socially considerate of it in day-to-day situations" (Nijssen et al. 2020: 332).

In a more recent paper, Spaccatini et al. (2023) examined how anthropomorphism in robots affects the way in which humans perceive and interact with other human beings. In particular, they wondered whether the attribution of the mind to anthropomorphic social robots impacts the empathic perception of other human beings. Their result indicates that "adopting social schemas with social robot is not a neutral process, but, rather, it affects the way we perceive other human beings subsequently. [...] although anthropomorphism is an advantage for Human–Technology Interactions, its possible consequent effect of how we perceive human beings cannot be neglected" (Spaccatini et al. 2023). According to their findings, anthropomorphic choices in robot design might have a mixed impact regarding empathic responses to other human beings. They could be positive, but what is more important in the context of this chapter is that they could also be negative. The authors recognized that their studies have limitations and that there is a need for more research, but they underscored the potential social consequences of their findings. Human-like robots might impact the way in which we treat other human beings, sometimes unfavorably.

Elsewhere, I wrote that hesitation in destroying robots in order to save human beings could be considered to be a crime (see, e.g., Mamak 2021). In some legal systems, there is a legal duty to help people who are in peril. Not taking action—like not pushing a robot in order to save humans—could be considered as not fulfilling the duty to rescue. This is sometimes referred to as Good Samaritan laws (see, e.g., McIntyre 1994; Feldbrugge 1965; Pardun 1997) that could lead to criminal charges.

Taking into consideration those two threats—epistemological and patient—it could be said that the human likeness of robots is not irrelevant from the perspective of human safety. In other contexts, these threats will pose risks of a different nature. There could be a domain in which these risks will have a limited meaning and be ignored as irrelevant while in another they could be a seriously dangerous design choice.

In a paper with Kaja Kowalczewska, 'Military robots should not look like a humans', we discussed these threats in a military setting (Mamak and

Kowalczewska 2023). We pointed out that deployment of robots in the military is problematic at many levels (ethical, legal, and social). If there is a decision to use robots there, the idea of human life creates additional issues. We pointed out that soldiers often need to make decisions about eliminating the enemy in a rush and, additionally, from a distance. If soldiers know that the other side is deploying human-like robots, then it is more dangerous to confront such entities face to face, and it is more rational to avoid such direct confrontation by eliminating potential problems as soon as possible, which increases the chances of mistakes and eliminating human beings instead. In the case of patient threat, we wrote that it is dangerous not only when robots are on the other side, but also when such robots are members of the same team. Attachment to robots and, for example, even a short hesitation to leave such robots on the battlefield could lead to someone's death. We concluded that, "While outside of the military context, it is not something obviously bad, in the military context, it brings additional risks to humans, who may not be rescued or who may lose their life to saving robots. We recommend not building robots that look like humans" (Mamak and Kowalczewska 2023: 7).

Indirect Risks

Indirect risks are focused on the long-term consequences of the deployment of humanoid robots on human safety. This section is particularly concerned with how communicating with human-like robots daily might decrease human safety. Two risks will need to be covered. First, there is a risk of changing the patterns of behaviors that could be harmful to humans; second, there is a risk of devaluing human-like shapes in public spaces.

Risk of Changing Patterns of Behavior toward Hmans

The term "moral patient" was mentioned earlier as referring to the entity that can be wronged. There is an ongoing discussion on whether robots could be moral patients, with strongly contradicting conclusions. (For an overview, see Gunkel 2018.) However, there is no similar doubt concerning the moral status of humans. Humans could be wronged, and the process of socialization aims to develop a set of skills that allow us to interact with fellow humans in a way that avoids mutual harm. For example, we need to know that while kicking a ball is acceptable, the same behavior performed on other human beings is not. In the literature about the societal consequences of deploying robots, there are indications that interactions with robots might change the ways in which we treat human beings, which could lead to different harms on the human side.

Vallor points out that technology has the potential to impact our moral practices (Vallor 2015; see also Vallor 2016). The impact could be positive or negative, which is her main concern. She worries that some technologies could lead to moral deskilling. She underscores that such an observation should be transferred into action, and technologies need to be developed with the awareness of the impact on moral life. In particular, she uses examples of the deskilling potential of social robots that could be adopted in care institutions.

In the discussion that could be classified as robot rights, one argument that is raised is as follows: violence against robots is problematic because it could lead to violence against humans. Whitby argues that the mistreatment of human-like robots is problematic because "those people who abuse human-like artifacts are thereby more likely to abuse humans" (Whitby 2008: 329). He based this claim on the analogy with computer games, which he suggests might be responsible for the increase in violence. More recent studies do not confirm that playing violent video games increases the violent behaviors of players (see, e.g., Kühn et al. 2019; Hilgard et al. 2019). Those findings do not mean that violence against robots will not affect the aggressive behaviors of humans, but rather that an analogy with violent computer games is not good for making recommendations regarding robots. There is a related discussion on that topic in Chapter 5.

The ethical discussion on sex robots also contains deliberations concerning the impact of the deployment of this kind of robot on human safety. There are objections regarding sex robots in general and those that are limited to more specialized sex robots, those which simulate rape and child-like sex robots. The most prominent representative of the movement against all sex robots is Kathleen Richardson. One of her arguments against sex robots is their potential for objectivization of women (Richardson 2015). The strong proposition to ban all sex robots was met with criticism that pointed out that it is not sufficiently justified (Hancock 2020; Danaher et al. 2017). In critique, it is often pointed out that sex robots might have some positive effects, for example, providing sexual pleasure to those who cannot achieve it due to their age, their state of illness, or place of stay (such as, military base) (see, e.g., McArthur 2017; Bendel 2021; Nucci 2017; Fosch-Villaronga and Poulsen 2020). The general ban of all sex robots would make it impossible to achieve any useful aims. However, the objection to banning all sex robots does not invalidate the worries that sex robots might also have a negative impact on the behaviors of humans. For example, Ozturk states: "Given the empathetic responses to abuse of such robots and the intuition that torturing them is wrong, treating robots in morally questionable ways is likely to create a perception that it is acceptable to treat humans— or at least other living beings—in the same way. Thus, it is imperative for the legal orders to act against the abuse of robots" (Ozturk 2020: 502; see also Bisconti Lucidi and Piermattei 2020: 557). One topic of concern regarding sex robots is the concept of consent. It is a problem for sex robots that simulate rape, but also sex robots in general (Frank and Nyholm 2017; Gutiu 2016). There is some concern that the presence of sex robots, who are always ready for sex, might create an unrealistic image of sexual interaction with human beings. Humans could be harmed if the same standards that were learned with interactions with robots were applied (see, e.g., Sparrow 2017).

The idea that supposedly would help to prevent associations that partners are always ready for sex is to install consent modules in sex robots (Peeters and Haselager 2021). When such a module was installed, sex robots would refuse sex from time to time. Elsewhere, I wrote that this idea, despite its good intentions, has drawbacks and could expose rape as an option for those who never thought about such an option before. In that sense, a proposed solution could be a hidden "rape module" (Mamak 2023: 81).

The discussion about possible long-term negative effects of the interaction with humanoid robots on humans has so far been based on speculations (see, e.g., Rothstein et al. 2021). There has not yet been an opportunity to conduct research on the lasting effects of abusive behaviors on robots and its impact on humans. However, there are pressing questions about whether we should allow or ban rape robots and child-like sex robots in light of the absence of empirical grounding. It is argued that despite uncertainty about their actual impact on the behaviors of users, they should be criminalized. The justification is not based on empirical grounds, but on other justifications such as legal moralism (see, e.g., Danaher 2017; Chatterjee 2020; Strikwerda 2017). What is more, Danaher doubts whether we should experiment with child-like sex robots at all when seeking the empirical base for regulation. He noted that the finding of conclusive and unambiguous answers to the question of whether such robots decrease or increase antisocial behaviors might be impossible in the near future, or ever (Danaher 2019).

Finding the ultimate answer to the question of whether the presence of human-like robots might lead to a decline in the safety of humans will not be easy in the future, and it is not currently possible to formulate this. Worries of this kind are formulated while discussing the ethical and societal consequences of deploying different kinds of robots. At least we should treat those worries seriously and be sensitive to any signs of a shift in human behaviors toward other humans due to interactions with human-like robots.

Risk of Devaluation of Human-like Shape

The last specific problem described in this chapter is the risk of devaluing human shape in social contexts. So far, among entities that we pass every day in social contacts, because only humans have human shape, we should treat human-shaped objects in a way that we treat humans. When we see an entity that looks like a human being and that requires help, our obligation should be to provide that assistance. As mentioned earlier, there is a moral duty to rescue humans who are in peril (see, e.g., Miller 2020; Ripstein 2000). In some countries, this duty has a legal character, and failing to fulfill it could lead to criminal charges being brought against such a person. There is a risk that with their presence in social life, entities that share human characteristics will lead to a reduction in the significance of human-shaped objects. More or less, we know how human organisms function. If we notice that a person is sleeping on a bench and we know that the night could be cold, we should wake up that person; otherwise, that person might freeze to death. When we drive a car and notice a human-shaped object in a trench, it is not likely that the person is just resting, but rather someone is in a situation that requires intervention. Due to the extent to which we value human life, we were required to form a detection mechanism that is sensitive to human harm. To some extent, this can be realized through a visual evaluation of the surroundings. These aspects of our functioning might be distorted by populating the social world with human-shaped entities that are not humans.

If we experience robots malfunctioning or being out of power, then the human-shaped object that does not move and stops working in unusual places might not be surprising or worrying. It raises the possibility that instead of determining whether a human requires assistance, we might assume that it is simply a robot that has stopped working. The presence of humanoid robots in public spheres might desensitize people to the view of human-like-shaped objects ("I think that it is a broken robot.")

This is the case in the fourth scenario outlined earlier in this chapter. The person who thought that the "person" was a robot and did not have malevolent intentions was devastated after realizing that they had been mistaken in their evaluation of the situation, but it was too to late do anything; the harm had been done.

What is described here could be seen as a variation of an epistemological threat, but it is not limited to that. In indirect threats, the focus is on the long-term consequences of the presence of human-like robots. The coexistence of human-shaped robots, due to the limited opportunity to know the true nature of the entities around us, combined with being able to bump into malfunctioning human-like robots, could dull the ability to detect humans needing help.

Conclusions

This chapter focused on safety issues related to creating robots that resemble humans in their external form. So far, the human-shaped entities that we spot in our everyday situations are humans. We learned about how we should treat fellow humans to ensure that we will not be harmful to each other. The arrival of human-shaped nonhuman entities might be socially disruptive, with drawbacks to human safety. The potential risks were divided into two categories: direct and indirect. Direct risks are involved with the impact of the physical presence of human-like robots in the vicinity of humans. We could simply mistake them for humans in dramatic situations due to our limited ability to gather information about the external world (epistemological threat). Knowing we had to deal with robots, we could also treat them in dilemmatic circumstances as entities deserving of moral consideration that dragged human beings into danger (patient threat). Indirect threats concern the long-term consequences of communicating with human-like robots. Through interactions with humanoid robots, we could change the standards of interacting with humans, for the worse. There is also the potential for devaluing the human-like shape in public spaces. So far, when we see a human-like object in an atypical state, we are morally and sometimes legally obliged to find out whether it is a human that may need help. In a world in which we could get used to the presence of robots that might malfunction at random moments, the view of human shapes that are in unusual places would not trigger concerns about human life.

References

Ashworth, A. J. 1975. "Self-Defence and the Right to Life." *The Cambridge Law Journal* 34 (2): 282–307.

Bartneck, Christoph, and Merel Keijsers. 2020. "The Morality of Abusing a Robot." *Paladyn, Journal of Behavioral Robotics* 11 (1): 271–83. https://doi.org/10.1515/pjbr-2020-0017.

Bendel, Oliver. 2021. "Love Dolls and Sex Robots in Unproven and Unexplored Fields of Application." *Paladyn, Journal of Behavioral Robotics* 12 (1): 1–12. https://doi.org/10.1515/pjbr-2021-0004.

Bertolini, Andrea. 2013. "Robots as Products: The Case for a Realistic Analysis of Robotic Applications and Liability Rules." *Law, Innovation and Technology* 5 (2): 214–47. https://doi.org/10.5235/17579961.5.2.214.

Bisconti Lucidi, Piercosma, and Susanna Piermattei. 2020. "Sexual Robots: The Social-Relational Approach and the Concept of Subjective Reference." In *Human-Computer Interaction. Multimodal and Natural Interaction*, edited by Masaaki Kurosu, 549–59. Lecture Notes in Computer Science. Cham: Springer International Publishing. https://doi.org/10.1007/978-3-030-49062-1_37.

Carlson, Zachary, Louise Lemmon, MacCallister Higgins, David Frank, Roya Salek Shahrezaie, and David Feil-Seifer. 2019. "Perceived Mistreatment and Emotional Capability Following Aggressive Treatment of Robots and Computers." *International Journal of Social Robotics* 11 (5): 727–39. https://doi.org/10.1007/s12369-019-00599-8.

Chalmers, David J. 1996. *The Conscious Mind: In Search of a Fundamental Theory*. New York.

Chatterjee, Bela Bonita. 2020. "Child Sex Dolls and Robots: Challenging the Boundaries of the Child Protection Framework." *International Review of Law, Computers & Technology* 34 (1): 22–43. https://doi.org/10.1080/13600869.2019.1600870.

Danaher, John. 2017. "Robotic Rape and Robotic Child Sexual Abuse: Should They Be Criminalised?" *Criminal Law and Philosophy* 11 (1): 71–95. https://doi.org/10.1007/s11572-014-9362-x.

———. 2019. "Regulating Child Sex Robots: Restriction or Experimentation?" *Medical Law Review* 27 (4): 553–75.

———. 2020. "Welcoming Robots into the Moral Circle: A Defence of Ethical Behaviourism." *Science and Engineering Ethic*. 26: 2023–49. https://doi.org/10.1007/s11948-019-00119-x.

Danaher, John, Brian Earp, and Anders Sandberg. 2017. "Should We Campaign Against Sex Robots?" In *Robot Sex: Social and Ethical Implications*, edited by John Danaher and Neil McArthur. MIT Press. https://mitpress.universitypressscholarship.com/view/10.7551/mitpress/9780262036689.001.0001/upso-9780262036689-chapter-004.

Feldbrugge, F. J. M. 1965. "Good and Bad Samaritans. A Comparative Survey of Criminal Law Provisions Concerning Failure to Rescue." *The American Journal of Comparative Law* 14 (4): 630–57. https://doi.org/10.2307/838914.

Fosch-Villaronga, Eduard. 2019. *Robots, Healthcare, and the Law. Regulating Automation in Personal Care*. London: Routledge. https://doi.org/10.4324/9780429021930.

Fosch-Villaronga, Eduard, and Adam Poulsen. 2020. "Sex Care Robots: Exploring the Potential Use of Sexual Robot Technologies for Disabled and Elder Care." *Paladyn, Journal of Behavioral Robotics* 11 (1): 1–18. https://doi.org/10.1515/pjbr-2020-0001.

Frank, Lily, and Sven Nyholm. 2017. "Robot Sex and Consent: Is Consent to Sex between a Robot and a Human Conceivable, Possible, and Desirable?" *Artificial Intelligence and Law* 25 (3): 305–23. https://doi.org/10.1007/s10506-017-9212-y.

Gunkel, David J. 2018. *Robot Rights*. Cambridge, Massachusetts: MIT Press.

———. 2023. *Person, Thing, Robot: A Moral and Legal Ontology for the 21st Century and Beyond*. Cambridge, Massachusetts: MIT Press.

Gutiu, Sinziana M. 2016. "The Roboticization of Consent." In *Robot Law*, edited by Ryan Calo, A. Froomkin, and Ian Kerr, 186–212. Edward Elgar Publishing. https://www.elgaronline.com/view/edcoll/9781783476725/9781783476725.00016.xml.

Hancock, Eleanor. 2020. "Should Society Accept Sex Robots: Changing My Perspective on Sex Robots through Researching the Future of Intimacy." *Paladyn, Journal of Behavioral Robotics* 11 (1): 428–42. https://doi.org/10.1515/pjbr-2020-0025.

Hansen, Luke R. 2023. "On the Existence of Robot Zombies and Our Ethical Obligations to AI Systems." *Journal of Social Computing* 4 (4): 270–74. https://doi.org/10.23919/JSC.2023.0023.

Heyman, Steven J. 1994. "Foundations of the Duty to Rescue." *Vanderbilt Law Review* 47 (3): 673–756.

Hilgard, Joseph, Christopher R. Engelhardt, Jeffrey N. Rouder, Ines L. Segert, and Bruce D. Bartholow. 2019. "Null Effects of Game Violence, Game Difficulty, and 2D:4D Digit Ratio on Aggressive Behavior." *Psychological Science* 30 (4): 606–16. https://doi.org/10.1177/0956797619829688.

Kant, Immanuel. 1991. *The Metaphysics of Morals*. Translated by Mary Gregor. Cambridge University Press.

Kirk, Robert. 1974. "Sentience and Behaviour." *Mind* 83 (329): 43–60.

———. 2021. "Zombies." In *The Stanford Encyclopedia of Philosophy*, edited by Edward N. Zalta, Spring 2021. Metaphysics Research Lab, Stanford University. https://plato.stanford.edu/archives/spr2021/entries/zombies/.

Krach, Sören, Frank Hegel, Britta Wrede, Gerhard Sagerer, Ferdinand Binkofski, and Tilo Kircher. 2008. "Can Machines Think? Interaction and Perspective Taking with Robots Investigated via fMRI." *PLOS ONE* 3 (7): e2597. https://doi.org/10.1371/journal.pone.0002597.

Kühn, Simone, Dimitrij Tycho Kugler, Katharina Schmalen, Markus Weichenberger, Charlotte Witt, and Jürgen Gallinat. 2019. "Does Playing Violent Video Games Cause Aggression? A Longitudinal Intervention Study." *Molecular Psychiatry* 24 (8): 1220–34. https://doi.org/10.1038/s41380-018-0031-7.

Levy, Neil, and Julian Savulescu. 2009. "Moral Significance of Phenomenal Consciousness." *Progress in Brain Researc*, 177: 361–70. https://doi.org/10.1016/S0079-6123(09)17725-7.

Malinowska, Joanna K. 2021a. "Can I Feel Your Pain? The Biological and Socio-Cognitive Factors Shaping People's Empathy with Social Robots." *International Journal of Social Robotic*y 14: 341–55. https://doi.org/10.1007/s12369-021-00787-5.

———. 2021b. "What Does It Mean to Empathise with a Robot?" *Minds and Machines* 31 (3): 361–76. https://doi.org/10.1007/s11023-021-09558-7.

Mamak, Kamil. 2021. "Whether to Save a Robot or a Human: On the Ethical and Legal Limits of Protections for Robots." *Frontiers in Robotics and AI* 8. https://doi.org/10.3389/frobt.2021.712427.

———. 2023. *Robotics, AI and Criminal Law: Crimes against Robots*. Routledge.

———. 2024. "Challenges of the Legal Protection of Human Lives in the Time of Anthropomorphic Robots." In *Cambridge Handbook on Law, Policy, and Regulations for Human-Robot Interaction*, edited by Woodrow Barfield, Yueh-Hsuan Weng, and Ugo Pagallo. Cambridge University Press.

Mamak, Kamil, and Kaja Kowalczewska. 2023. "Military Robots Should Not LooklLike a Humans." *Ethics and Information Technology* 25. https://doi.org/10.1007/s10676-023-09718-6.

McArthur, Neil. 2017. "The Case for Sexbots." In *Robot Sex: Social and Ethical Implications*, edited by John Danaher and Neil McArthur. Cambridge, MA: MIT Press. https://mitpress.universitypressscholarship.com/view/10.7551/mitpress/9780262036689.001.0001/upso-9780262036689-chapter-004.

McIntyrE, Alison. 1994. "Guilty Bystanders? On the Legitimacy of Duty to Rescue Statutes." *Philosophy & Public Affairs* 23 (2): 157–91. https://doi.org/10.1111/j.1088-4963.1994.tb00009.x.

Miller, David. 2020. "The Nature and Limits of the Duty of Rescue." *Journal of Moral Philosophy* 17 (3): 320–41. https://doi.org/10.1163/17455243-01703003.

Nielsen, Yngwie Asbjørn, Stefan Pfattheicher, and Merel Keijsers. 2022. "Prosocial Behavior toward Machines." *Current Opinion in Psychology* 43 (February): 260–65. https://doi.org/10.1016/j.copsyc.2021.08.004.

Nijssen, Sari R. R., Barbara C. N. Müller, Rick B. van Baaren, and Markus Paulus. 2019. "Saving the Robot or the Human? Robots Who Feel Deserve Moral Care." *Social Cognition* 37 (1): 41–52. https://doi.org/10.1521/soco.2019.37.1.41.

Nijssen, Sari R. R., Evelien Heyselaar, Barbara C. N. Müller, and Tibor Bosse. 2020. "Do We Take a Robot's Needs into Account? The Effect of Humanization on Prosocial Considerations Toward Other Human Beings and Robots." *Cyberpsychology, Behavior, and Social Networkin*, 24 (5r. https://doi.org/10.1089/cyber.2020.0035.

Nucci, Ezio Di. 2017. "Sex Robots and the Rights of the Disabled." In *Robot Sex: Social and Ethical Implications*, edited by John Danaher and Neil McArthu0. Cambridge, MA: MIT Press. https://doi.org/10.7551/mitpress/9780262036689.003.0005.

Ozturk, Anil. 2020. "Anthropomorphic Machines: Implications of Human-Robot Social Interactions for Law and Society." International Scientific Conference:"*Transformative Technologies: Legal and Ethical Challenges of the 21st Century*" January. https://www.academia.edu/49786300/Anthropomorphic_Machines_Implications_of_Human_Robot_Social_Interactions_for_Law_and_Society.

Pardun, John T. 1997. "Good Samaritan Laws: A Global Perspective Comment." *Loyola of Los Angeles International and Comparative Law Journal* 20 (3): 591–614.

Peeters, Anco, and Pim Haselager. 2021. "Designing Virtuous Sex Robots." *International Journal of Social Robotics* 13 (1): 55–66. https://doi.org/10.1007/s12369-019-00592-1. https://doi.org/10.11647/obp.0366.01.

Prescott, Tony J. 2017. "Robots Are Not Just Tools." *Connection Science* 29 (2): 142–49. https://doi.org/10.1080/09540091.2017.1279125.

Richardson, Kathleen. 2015. "The Asymmetrical 'Relationship.'" *ACM SIGCAS Computers and Society* 45 (3): 290–93. https://doi.org/10.1145/2874239.2874281.

Riek, Laurel D., Tal-Chen Rabinowitch, Bhismadev Chakrabarti, and Peter Robinson. 2009. "How Anthropomorphism Affects Empathy Toward Robots." In *Proceedings of the 4th ACM/IEEE International Conference on Human Robot Interaction*, 245–46. HRI '09. New York, NY, USA: ACM. https://doi.org/10.1145/1514095.1514158.

Ripley, David L., and Thomas Politzer. 2010. "Vision Disturbance after TBI." *NeuroRehabilitation* 27 (3): 215–16. https://doi.org/10.3233/nre-2010-0599.

Ripstein, Arthur. 2000. "Three Duties to Rescue: Moral, Civil, and Criminal Special Issue: The Moral and Legal Limits of Samaritan Duties." *Law and Philosophy* 19 (6): 751–80.

Rosenthal-von der Pütten, Astrid M., Frank P. Schulte, Sabrina C. Eimler, Sabrina Sobieraj, Laura Hoffmann, Stefan Maderwald, Matthias Brand, and Nicole C. Krämer. 2014. "Investigations on Empathy towards Humans and Robots Using fMRI." *Computers in Human Behavior* 33 (April): 201–12. https://doi.org/10.1016/j.chb.2014.01.004.

Rosenthal-von der Pütten, Astrid M., Nicole C. Krämer, Laura Hoffmann, Sabrina Sobieraj, and Sabrina C. Eimler. 2013. "An Experimental Study on Emotional Reactions Towards a Robot." *International Journal of Social Robotics* 5 (1): 17–34. https://doi.org/10.1007/s12369-012-0173-8.

Rothstein, Nina J., Dalton H. Connolly, Ewart J. de Visser, and Elizabeth Phillips. 2021. "Perceptions of Infidelity with Sex Robots." In *Proceedings of the 2021 ACM/IEEE International Conference on Human-Robot Interaction*, 129–39. HRI '21. New York, NY, USA: Association for Computing Machinery. https://doi.org/10.1145/3434073.3444653.

Salvini, P., G. Ciaravella, W. Yu, G. Ferri, A. Manzi, B. Mazzolai, C. Laschi, S. R. Oh, and P. Dario. 2010. "How Safe Are Service Robots in Urban Environments? Bullying a Robot." In *19th International Symposium in Robot and Human Interactive Communication*, 1–7. https://doi.org/10.1109/ROMAN.2010.5654677.

Sanoubari, Elaheh, James Young, Andrew Houston, and Kerstin Dautenhahn. 2021. "Can Robots Be Bullied? A Crowdsourced Feasibility Study for Using Social Robots in Anti-Bullying Interventions." In *2021 30th IEEE International Conference on Robot & Human Interactive Communication (RO-MAN)*, 931–38. https://doi.org/10.1109/RO-MAN50785.2021.9515450.

Spaccatini, Federica, Giulia Corlito, and Simona Sacchi. 2023. "New Dyads? The Effect of Social Robots' Anthropomorphization on Empathy towards Human Beings." *Computers in Human Behavior* 146 (September): 107821. https://doi.org/10.1016/j.chb.2023.107821.

Sparrow, Robert. 2017. "Robots, Rape, and Representation." *International Journal of Social Robotics* 9 (4): 465–77. https://doi.org/10.1007/s12369-017-0413-z.

Strikwerda, Litska. 2017. "Legal and Moral Implications of Child Sex Robots." In *Robot Sex: Social and Ethical Implications*, edited by John Danaher and Neil McArthu0. Cambridge, MA: MIT Press. https://doi.org/10.7551/mitpress/9780262036689.003.0008.

Suzuki, Yutaka, Lisa Galli, Ayaka Ikeda, Shoji Itakura, and Michiteru Kitazaki. 2015. "Measuring Empathy for Human and Robot Hand Pain Using Electroencephalography." *Scientific Reports* 5 (1): 15924. https://doi.org/10.1038/srep15924.

Vallor, Shannon. 2015. "Moral Deskilling and Upskilling in a New Machine Age: Reflections on the Ambiguous Future of Character." *Philosophy & Technology* 28 (1): 107–24. https://doi.org/10.1007/s13347-014-0156-9.

———. 2016. *Technology and the Virtues: A Philosophical Guide to a Future Worth Wanting*. 1st edition. New York, NY: Oxford University Press.

van de Poel, Ibo, Jeroen Hopster, Guido Löhr, Elena Ziliotti, Stefan Buijsman, and Philip Brey. 2023. "1: Introduction." In *Ethics of Socially Disruptive Technologies*, edited by Ibo van de Poel, Lily Eva Frank, Juli Hermann, Jeroen Hopster, Dominic Lenzi, Sven Nyholm, Behnam Taebi, and Elena Ziliotti, 11–32.

Veruggio, Gianmarco, Fiorella Operto, and George Bekey. 2016. "Roboethics: Social and Ethical Implications." In *Springer Handbook of Robotics*, edited by Bruno Siciliano and Oussama Khatib, 2135–60. Springer Handbooks. Cham: Springer International Publishing. https://doi.org/10.1007/978-3-319-32552-1_80.

Wallach, W., and C. Allen. 2009. *Moral Machines: Teaching Robots Right from Wrong*. Oxford University Press.

Whitby, Blay. 2008. "Sometimes It's Hard to Be a Robot: A Call for Action on the Ethics of Abusing Artificial Agents." *Interacting with Computers* 20 (3): 326–33. https://doi.org/10.1016/j.intcom.2008.02.002.

Wynsberghe, Aimee van. 2016. *Healthcare Robots: Ethics, Design and Implementation*. Abingdon, Oxon: Routledge.

Złotowski, Jakub, Hidenobu Sumioka, Shuichi Nishio, Dylan F. Glas, Christoph Bartneck, and Hiroshi Ishiguro. 2016. "Appearance of a Robot Affects the Impact of Its Behaviour on Perceived Trustworthiness and Empathy." *Paladyn, Journal of Behavioral Robotics* 7 (1): 55–66. https://doi.org/10.1515/pjbr-2016-0005.

3 | Representational Issues of Human-like Robots

Introduction

The decision to make robots human-like raises representational issues. We see in humanoid robots not mere resemblance of our species, but also a range of characteristics that are ascribed to humans, like gender, race, and age. The human likeness of robots is crucial to eliciting their social categorization (Perugia et al. 2022). Robots could represent individuals and specific groups of people. Giving human-like robots the characteristics of a certain kind could be beneficial from the perspective of usability. Playing with characteristics that resemble specific group members in the design of human-like robots could be used to promote positive interactions with robots and favorable attributes for interactions with them (Złotowski et al. 2015: 354). Some groups could accept that and even be proud that some robots are similar to them, but this will not always be the case. As the issue of ethics usually focuses on the right and wrong, the focus of this chapter too is on those scenarios that could go wrong and have negative effects. Groups and individuals that are misrepresented or not represented at all could suffer real-life negative consequences from design choices. The negative impact could be those that are not intended at all by the designers; some results are side-effects of choices that aimed to bring positive effects to robots' functioning. Making the association between robots and social categories is unavoidable to some extent; as Coeckelbergh points out that in the case of social robots, there are always some links to society (Coeckelbergh 2022: 74).

I want to illustrate these problems with a few scenarios. In the first scenario, a young singer is gaining popularity due to a couple of hits from her debut album. She needs to get used to the fact that many people are interested in her life. The media sources outdo each other in their attempts to publish new information about her life. She has a feeling that her face is everywhere. At the same time, a new robotic company is planning to launch its latest commercial product, companion robots. They believe it could have the potential to be the first mass-produced humanoid companion robot. They are planning a marketing strategy, and someone in the company proposes that they use the popularity of that young star. They believe they could save on marketing costs and get free media coverage. Without asking for permission, they create a robot resembling this singer. The robot has the same size, body parameters, and hair, and has a similar voice tone, but some minor facial details are different. Despite those slight differences, the typical observer immediately associates the new robot with the singer. When a product is ready, the robot company starts its campaign with videos. Observers think at first that it is a human, but then it is revealed that it is a robot that could be bought immediately. The company executives were right; the product became a sales hit, partially because it resembled the popular star. Initially, the star ignores the robot, but then she starts seeing trending videos on social media platforms with "her" in weird contexts. She is tagged in those videos. The media starts to ask her for explanations regarding things she allegedly did or said. They struggle to recognize which movie is with her, and which is with robots that resemble her. The star is devastated, as is her image.

In the second scenario, a company's brand-new medical humanoid robot is becoming better known as a result of a government contract. This company is gaining a competitive edge over the others and is present in a growing number of places. Due to its access to extensive resources, it can constantly improve its products and enhance its market position. The designers decide that their robots should look like White males. The decision is a result of several reasons. The male characteristics were chosen according to the research which showed that robots with male features were perceived as more competent than the same robots with female features. The research was conducted partially by one of the team members who transferred from academia to industry. Whiteness, in turn, was not a matter of deliberate choice nor of in-depth discussion. It was a natural choice for the whole group of designers and engineers, who were all White people. The robot became popular. It was sent to rural places that do not have human doctors. For the first time, some communities had access to medical care, in which human medical robots were the first (and sometimes the only) face of the medical world. As a result, some hospitals reported cases of patients who for years had contact only with humanoid medical robots and who, after getting to a hospital, asked for a male doctor, and refused to be examined by a female doctor because they were afraid that a female doctor would be unable to help them. Robots are also present in popular culture. There is a hit TV series with a couple of seasons in which a robot doctor solves the most complex medical puzzles. Building on the wave of popularity, medical schools observed an increased number of study candidates. However, the increase was observable in one social group—White men, who identify with the star of the show.

The last-mentioned scenario is happening in a country that is divided almost in half in terms of religious affiliation. The representatives of one group gain power in the election, and due to a series of manipulations in public media, by spending public money, and by suppressing democratic institutions, they remain in power for a few decades. In the meantime, the era of humanoid robots has begun. Neighboring countries started to buy robots for public purposes. Observing the expenditure of public money on robots, people associated with the government became aware of the opportunity to earn extra money on contracts for the supply of robots for institutions, such as police or fire departments. New robots commissioned by the government need to have certain characteristics. Besides having typical functionalities, there are specific requirements regarding the external design of robots. They need to be equipped with the symbols of religion associated with the government. The robots in public spaces use religious greetings and have religious symbols visible to all interacting with them. The presence of such robots, which become visible in all public places, deepened the feeling of not belonging to the country among those citizens whose religious beliefs were different. What is more, police robots started to be used in riots. They were allowed to use force. Although robots, not humans, use violence, the religious symbols make an association with the ruling party and, indirectly, with their religion, which strengthens mutual aversion in society.

These examples show many aspects that are related to the representative power of human-like robots. In this chapter, I have tried to write about these issues in a more systematic way. This chapter is structured as follows. After the introductory part, the focus is on how robots are socially categorized. The discussion then moves to risks related to representation. After that, representational humans are confronted with the call for making robots beneficial for society. I then briefly present ideas on how to limit negative consequences related to representation. The final section concludes.

Social Categorization of Humanoid Robots

Human-like robots do not just look like humans, but they usually look like humans belonging to certain social categories. For example, most robots are perceived as young adults (Perugia et al. 2022). That age impression refers not to the number of actual years of robots but to the age range of humans that is reflected in the shape of the robot's body. There are more social characteristics that are perceivable in humans, but at the same time, there is a lack of a full range of options in robots' looks. Riek and Howard point out that there is a lack of diversity in robot morphology and design. They note that robots represent only a fraction of the possible varieties (Riek and Howard 2014). For example, robots usually reflect the body of a human with all fully working body parts. Some in that context are talking about the problem of ableism, which is discrimination against people with disabilities. (On ableism and technology, see Shew 2020). It is pointed out that almost always bodies of robots refer to "fully-abled bodies" (Zaga et al. 2022: 26).

It must be noted that when there is a reference to different social categories, it is based on how those categories are perceived in robots' design by those who interact with robots and not how those categories should be treated in society at large. Those perceptions that are described in the studies in the field of human–robot interactions are often based on simplifications. (For more nuanced views, see, e.g., Briggs and George 2023; Malinowska and Żuradzki 2022.) The reason to focus on them is to illustrate the potential harm that could occur when categorization is used by creators of robots or when categorization is done by the people who interact with robots.

One of the social categories that is heavily discussed in the context of robot design is gender. This is also the most examined issue in the experimental settings compared to other possible social characteristics. Perugia and Lisy, in a review paper on gender and robots in human–robots interaction field, identified 553 relevant papers (Perugia and Lisy 2023). It is a number that far exceeds the number of papers in any other category (e.g., age or race). Gendering in robots is a good example to show the complexity of the problem and how it is unavoidable.

Gendering robots could be seen as a way to make robots more acceptable in some social contexts. Carpenter shows a preference for female-gendered robots in the home setting (Carpenter et al. 2009). Greater acceptance of female-gendered robots has also been shown in other studies in the healthcare context (Tay et al. 2014). However, male-gendered robots were considered to be more suitable for security-related tasks (Tay et al. 2013). Gender choices in design might impact the judgments about the competence of robots (Bernotat et al. 2021).

Gendering robots could be achieved by many different means, including the overall look, shape of the body, and clothes, naming a few of the more obvious. Gender might be associated with less obvious characteristics like walking style (Alesich and Rigby 2017: 52) or hair length (Eyssel and Hegel 2012). Gendering, as already mentioned, is to some extent unavoidable. For example, Søraa points out that the nature of language might promote the use of gendered pronouns when talking about robots (Søraa 2017). Roesler et al. in their studies have examined whether there is "a male-robot bias." They studied the role of language and anthropomorphism in gendering robots. They showed that "masculine grammatical gender tends to reinforce a male ascription of gender-neutral robots" (Roesler et al. 2023: 1). They point out that the "male-robot biases" are associated with both the appearance of most anthropomorphic robots and the language that is used when referring to the robot. In other words, even if the designers try to make the robot gender neutral, the male gender is likely to be perceived in that robot anyway.

Also, race, as a social category, is transferred to human-like robots. Humanoid robots are racialized and mostly presented as White (Cave and Dihal 2020). When robots are presented as belonging to different races, prejudices known from human–human interactions might be applied. Studies by Berneck et al. suggest that people extend racial prejudices from humans to robots that are portrayed as representing that category of humans (Bartneck et al. 2018). Barfield, in her studies, collected responses to the same robot that was colored in different ways, and the robot was perceived differently (suitable for certain jobs or likeliness of committing an assault) based on its color (Barfield 2021). In studies conducted by Strait et al., robots were presented as belonging to one of three categories—White, Asian, and Black. The authors collected responses to those robots and found out that people more frequently dehumanize robots racialized as Asian and Black than those presented as White (Strait et al. 2018). As with gender, race is also a category that is not easy to avoid. Sparrow explains that a robot perceived as White does not have white color; it could be silver or gold as well; whiteness is not about color (White people are not white in the literal sense). The race of robots could be derived not only from the color of robot bodies but also from specific characteristics like morphology, hair, clothes, accent, behavior, or social roles (Sparrow 2020b). Sparrow adds that the more human-like robots are, the easier it is to perceive them as having racial characteristics and that "robots do have race regardless of the intentions of their designers" (Sparrow 2020b: 545).

There might be social categories that are easier to avoid than gender or race. In the third scenario from the beginning of this chapter, there was mention of robots that could be seen as representatives of certain religions. The design choices to represent specific religions were made deliberately in that scenario. There was usage of voice patterns (religious greetings in interaction with humans) and visible religious symbols on the robots' bodies. Resignation from carrying a religious component could be easily achievable in the provided scenario. Usage of religiously neutral language and removing religious signs would end robots being seen as representatives of a given category. However, in some societies, avoiding the religious categorization of robots would not be easy. If, in each society, racial divisions are related to religious ones, then the racialization of robots could carry the religious component in it.

Representational Harms

One of the central ethical issues related to artificial intelligence (AI) is biases (see, e.g., Fazelpour and Danks 2021; Noble 2018; Buolamwini and Gebru 2018). It could be harmful to members of groups when the data that is used to make decisions is biased. AI tools that are biased could lead to unjust decisions in different life situations, like bank loan applications, job applications, or parole decisions. Biases in AI are also problematic in the context of robots. Robots that use biased machine learning models who act in the real world could "through bodies physically amplifying malignant stereotypes in general" (Hundt et al. 2022: 743). This could be especially dangerous in the case of robots that represent the state, such as police robots, that could enforce biased decisions in the wild (Mamak 2023: 127)

In the literature concerning biases in AI, there is used the concept of "representational harm" (cf. Wang et al. 2022; Abbasi et al. 2019; Katzman et al. 2023; Buddemeyer et al. 2021). Representational harm in the context of biases in AI usually is related to the problem of stereotyping (see more about race and technology in Benjamin 2019). According to Blodgett et al., "representational harms arise when a system (e.g., a search engine) represents some social groups in a less favorable light than others, demeans them, or fails to recognize their existence altogether" (Blodgett et al. 2020:

2). Shelby et al. have developed a list of subtypes of representational harms to social groups that include: stereotyping (some professions are not suitable for a woman); demeaning (making associations of crimes with certain races); erasing (minorities do not exist in some contexts); alienating (some groups do not belong to society/parts of society); denying people the opportunity to self-identify (apps using classifiers that use binary genders), and reifying essentialist social categories (system classifies social membership based on narrow criteria) (Shelby et al. 2023: 6). For example, comments by Poulsen et al. on satiation, in which an automatic system flags drag queens as toxic, point out that as a result of such practices, "queer people will remain mostly invisible, silent, powerless and unable to understand how these technologies may affect them" (Poulsen et al. 2020: 152). They believed that by framing technology with traditional categories of sex, minorities might be excluded.

Representational harm could also be a problem in the case of robots. When a specific characteristic is chosen to represent particular kinds of robots, it impacts the group that is being represented (or omitted) by such choices. Imagine the example of robots deployed to do extremely low-esteem jobs in a given society. Suppose those robots resemble a group of humans present in that society. In that case, the design choices may constitute representational harm to that group of people by creating an association with that task. Robots' bodies, via anthropomorphism, may bring representational harm that should be avoided.

As mentioned, robots might represent types of people and particular people (Sparrow 2020a: 144). Some people might want to be represented by robots. An example of such an attitude is the case of Japanese roboticist Hiroshi Ishiguro, who builds robot versions of himself (see, e.g., Brownlee 2007). However, representing individuals in robots is risky due to potential consequences. Robots' bodies could be used to bring representational harm to individuals. In the first scenario from the beginning of this chapter, the victim of such representation is a famous singer whose public image has been destroyed by deploying robots resembling her. Lancaster pointed out that creating sex robots that resemble particular individuals becomes easier. She worries that it could lead to the objectification of the person who is presented in such robots and believes that if it happens without the consent of such a person, it should be considered intrinsically wrong (Lancaster 2021).

The problem of representational harm for people who belong to social categories that suffer from biases does not lie directly in how robots are perceived but in how human–robot relations might impact human–human relations. Robots do not feel that they could be harmed; the usage of harmful stereotypes does not bother them, but we should be concerned with the problems due to the interests of humans (on the possibility of discriminating self-aware robots, see Barfield 2023). Dovidio et al., in their work about stereotypes, point out that stereotypes can "not only promote discrimination by systematically influencing perceptions, interpretations, and judgments, but they also arise from and are reinforced by discrimination, justifying disparities between groups" (Dovidio et al. 2010: 7). It might be the case of robots, as Spaccatini et al. note that the results of their experiment indicate "that adopting social schemas with social robot is not a neutral process, but, rather, it affects the way we perceive other human beings subsequently" (Spaccatini et al. 2023: 8). Coeckelbergh noted that in the context of personal robots, biases toward people could be created, sustained, or increased (Coeckelbergh 2022: 73). Robots, compared to other forms of representation (e.g., photos), could share more properties with an object that they represent, like size and shape, and that multiplicity of ways of representation raises distinctive ethical issues (Sparrow et al. 2023: 1716). The next section takes up the problem of representational harm with the concept of robots for social good. There will be a discussion on why representational issues should be a matter of political concern.

Robots for Social Good

There is an ongoing and almost omnipresent discussion about the impacts of AI on our lives and planet. Parts of that discussion focus on AI risks—those that are here and now, such as biases, issues with responsibility, and privacy concerns (cf. Bartneck et al. 2021; Coeckelbergh 2020). The fraction of risk-focused debate concerns more far-reaching, long-term consequences and the potential extensional risks related to AI (cf. Vold and Harris 2021; Müller and Cannon 2021). This part of the discussion is polarizing, and strong positions defended by high-profile experts are sometimes exclusive (cf. Nature 2023; Sætra and Danaher 2023). On the other end of the spectrum of the discussion on the impact of AI are topics focused on how AI could contribute to a better world. In the literature, the phrase "AI for social good" (or "AI for good") is in use. Floridi et al. note that there is still little understanding of what constitutes AI "for social good" and propose a definition:

"the design, development, and deployment of AI systems in ways that (i) prevent, mitigate or resolve problems adversely affecting human life and/or the wellbeing of the natural world, and/or (ii) enable socially preferable and/or environmentally sustainable developments" (Floridi et al. 2020: 1773–74).

The less-developed subproblem of "AI for social good" is the idea of "robots for social good" being concerned with the issue of how to ensure that robots will be of benefit to society (cf. Lee et al. 2019). However, what is considered "good" is relative, and it is helpful to be more definite. In the case of "AI for social good," it is also problematic what "good" means (see, e.g., Berendt 2019). Moore suggests that the field "AI for social good" should be better referred to as "AI for not bad" (Moore 2019). Tomašev et al., in their paper about AI and social good, link the concept with the United Nations' Sustainable Development Goals (SDGs) (Tomašev et al. 2020; see also Stahl et al. 2023). Cowls et al. propose using SDG as a benchmark for evaluating "AI for social good" tools (Cowls et al. 2021). There is an ongoing discussion on how AI is related to SDGs (cf. Sætra 2022, 2023; Mazzi and Floridi 2023). Now I want to link the design choices of robots with the matter of SDGs.

The starting point for this thinking is that technology is not politically neutral (cf. Ihde 1990; Verbeek 2011). Winner, in his piece entitled "Do Artifacts Have Politics?" affirmatively answers that title question and shows, with examples, how choices in technology may have a political meaning (Winner 1980). Sparrow asks a similar question in the piece "How Robots Have Politics" (Sparrow 2021), in line with Winner's position, arguing that robots have even more political meaning and become robots by combining the political aspects of computer tools with the physicality of robot bodies. How do the representational harms in robots link to SDGs? It was already pointed out that the harmful stereotypes that would be applied to robots could impact human–human interactions and deepen inequalities (see also Mamak et al. 2023). Inequalities are the matter of SDG goal 5, which is about "gender equality," and goal 10 that aims to "reduce inequalities." Deploying humanoid robots, knowing that robot design choices could worsen the situation of various social groups within a given society, is a political matter that should be somehow responded to. In the next section, some ideas on that matter will be discussed.

How Do You Respond to Representational Harms?

One of the more intuitive responses to the question of avoiding representational harm in robot design seems to be avoidance of human likeness in the first place, or at least avoidance of stereotyping in the design of robots (see, e.g., Dufour and Ehrwein Nihan 2016; Bryson 2018). In a paper entitled "Robotics Has a Race Problem," Sparrow pointed out the dilemma faced by designers of humanoid robots. He noted that most robots, by default, are white and are perceived as being White, which could expose those designers to accusations of racism. But if we would like to represent robots as belonging to different races, then the association of slavery might arise due to the service nature of robots. His proposition is to create robots in a way that will make people struggle to attribute specific races to them or to make robots that do not resemble humans (Sparrow 2020b). However, as mentioned earlier, and also noted by Sparrow, avoidance of the representational component is not easy to achieve. Perugia and Lisy, in a review paper about the general issues in robots, noted that gendering is not a process that is entirely controllable (Perugia and Lisy 2023: 3).

Resigning from the human likeness is an option, but in this book the focus is on the design of human-like robots. If our starting point is to build human-like robots, then users' social categorization seems unavoidable. So far, stereotyping in robotics has been presented as something that could admittedly be useful in the case of the usability of robots, but is linked to possible negative consequences for groups that are stereotyped. It has also been pointed out in the literature that stereotyping could be used to achieve positive effects. Eyssel and Hegel, while discussing the use of gender stereotypes in robot design, ask the question whether we should do that, and in response to that question, the authors present two alternatives. First, to use and explain them to manipulate users' mental models. The second alternative is to construct counter-stereotypical machines; they use the examples of female-gendered robots that perform tasks that are considered typically manly. The authors note that these "questions seem to be worth considering by developers and designers of robots because of their social and societal consequences" (Eyssel and Hegel 2012: 2224). The second option is seen as design choice that has the power to change some stereotypes. Using stereotyping for good purposes seems to have been suggested by Złotowski et al. in a paper about the challenges of anthropomorphism in robotics. The authors refer to empirical research that shows that positive contact with an outgroup can reduce prejudice or negative feelings about the group (Złotowski et al. 2015: 354). They wonder whether the same mechanism might work in the case of the interaction between humans and robots.

We can imagine situations in which smooth interaction with robots can be crucial. When robots are used in dangerous situations, in which there is no time for hesitation, the use of stereotypes might be defended to ensure that interaction with robots will be smooth and intuitive. But it seems that most humanoid robots would operate in less dramatic circumstances. Even if the initial counterintuitive and anti-stereotype look of robots makes it difficult to interact, the effect of such counteractive interaction could be worth the initial troubles in the long term. The initial acceptability of robots should not be the final argument in the choice of what a robot looks like in all possible areas of the application of robots.

It is worth going back to Sparrow, who links the idea of humanoid robots with the concept of slavery (Sparrow 2020b). When someone looks at the humanoid project from that perspective, any representation of humans in robots is risky and potentially degrading for humanity. However, there could be justified doubts about whether slavery is the framework that humans will adopt for thinking about all humanoid robots. For example, some brands use AI influencers for advertising. Studies show that such entities could have a positive impact on products, similar to their human counterparts (see, e.g., Thomas and Fowler 2021). There is also a discussion about love with robots (Levy 2008; Nyholm and Frank 2019; Sullins 2012; Sætra 2021; Mamak 2022). It is doubtful whether slave framing is appropriate to describe such relationships with robots, and whether people who claim to be in love with robots would refer to it as a slave at the same time.

Trying to combine what has been said so far in this section, taking into consideration the concerns presented by Sparrow and the positive potential of using stereotypes, it might be defended that in the case of professions that are treated in society with low esteem, the design of robots should not use human likeness at all, or at least should avoid stereotyping to the extent that it is achievable. However, in the case of professions that enjoy a positive status in society, the design can play more with the concept of stereotyping to promote positive values and diversity. In other words, when robots are used in contexts that could be perceived negatively, human likeness, or at least stereotyping, allowing robots to be linked to specific social groups, should be avoided. Where robots will be used in a more positive context, stereotypes can be more permissible. If usage in design reference to specific social groups is not necessary (e.g., for safety reasons), then the design of robots should be used to promote social good, equity, and diversity, limiting the harmful stereotypes in society. Using robot bodies for diversity is easier to say than do. Lupetti points out that "the design for diversity, equity and inclusion is a complex challenge that requires existing approaches and methods to be revised" (Lupetti 2022: 37)

Weßel et al. have discussed issues with gendering robots in the context of eldercare. They described three solutions to problems: (a) the explanation of robotic functions to dispel gender perspectives; (b) the neutralization of gender attributions; or (c) the queering of the attributions (Weßel et al. 2023: 1969). The authors doubt whether the two first strategies could work; for example, they point out that the explanation could not be sufficient, and—regarding the second strategy—even if we try to avoid gendered robots, it is likely that they will be gendered despite those efforts. In their proposition to queering robots, the authors see potential drawbacks, but believe that this idea could have the most promising potential: "Queering might be able to acknowledge the inevitable relevance of gender aspects in human–robot interaction while at the same time challenging their conventional and normalizing application and promotion" (Weßel et al. 2023: 1972).

In the context of using design for more diverse purposes, it is worth looking at the proposition by Williams, who discusses how the call for a more equitable design of robots could manifest itself, and lists three approaches. First, robots are designed by, with, and for a specific community. Second, community members could change their appearance and behavior. Third, a robot changes its appearance and behavior based on the community in which it happens to interact. In his paper, he focused on the last approach, underscoring that despite noble intentions, the problem is ethically complex and could bring some additional issues if done incorrectly (Williams 2023: 1). Williams' analysis illustrates how complex it is to employ a more diverse design.

Conclusions

To sum up, in this chapter, the representational aspects of human-like robots were discussed. Humans apply social categories to robots, seeing in them not just humans but humans that belong to social categories. Characteristics that are ascribed to robots include age, gender, or race, to name a few. People not only see those categories in robots but also behave towards robots as if they belong to social groups. This mechanism could be used to achieve better performance in human–robot interactions or higher acceptability of robots, but it is also associated with costs. The existing stereotypes in society could be strengthened by applying those patterns to robots. Human-like bodies in robots could pose representational risks. Robots with their bodies carry political baggage that may impact many groups in society. This political issue should not be ignored at the level of designing robots and should be a matter of concern to political bodies. There were also discussions about possible responses to representational risks with an emphasis on using stereotyping to build a better, more equal, and diverse society. However, there is no easy way to achieve that goal. Good intentions might not be enough, and the solutions might bring new problems. More work needs to be done, both empirical and theoretical, to approach representational issues in robot design.

References

Abbasi, Mohsen, Sorelle A. Friedler, Carlos Scheidegger, and Suresh Venkatasubramanian. 2019. "Fairness in Representation: Quantifying Stereotyping as a Representational Harm." In *Proceedings of the 2019 SIAM International Conference on Data Mining (SDM)*, 801–809. Proceedings. Society for Industrial and Applied Mathematics. https://doi.org/10.1137/1.9781611975673.90.

Alesich, Simone, and Michael Rigby. 2017. "Gendered Robots: Implications for Our Humanoid Future." *IEEE Technology and Society Magazine* 36 (2): 50–59. https://doi.org/10.1109/MTS.2017.2696598.

Barfield, Jessica K. 2021. "Discrimination and Stereotypical Responses to Robots as a Function of Robot Colorization." In *Adjunct Proceedings of the 29th ACM Conference on User Modeling, Adaptation and Personalization*, 109–14. UMAP '21. New York, NY: Association for Computing Machinery. https://doi.org/10.1145/3450614.3463411.

———. 2023. "Discrimination against Robots: Discussing the Ethics of Social Interactions and Who Is Harmed." *Paladyn, Journal of Behavioral Robotics* 14 (1): 20220113. https://doi.org/10.1515/pjbr-2022-0113.

Bartneck, Christoph, Christoph Lütge, Alan Wagner, and Sean Welsh. 2021. *An Introduction to Ethics in Robotics and AI*. SpringerBriefs in Ethics. Cham: Springer International Publishing. https://doi.org/10.1007/978-3-030-51110-4.

Bartneck, Christoph, Kumar Yogeeswaran, Qi Min Ser, Graeme Woodward, Robert Sparrow, Siheng Wang, and Friederike Eyssel. 2018. "Robots And Racism." In *Proceedings of the 2018 ACM/IEEE International Conference on Human–Robot Interaction*, 196–204. Chicago, IL: ACM. https://doi.org/10.1145/3171221.3171260.

Benjamin, Ruha. 2019. *Race after Technology: Abolitionist Tools for the New Jim Code*. 1st edition. Medford, MA: Polity.

Berendt, Bettina. 2019. "AI for the Common Good?! Pitfalls, Challenges, and Ethics Pen-Testing." *Paladyn, Journal of Behavioral Robotics* 10 (1): 44–65. https://doi.org/10.1515/pjbr-2019-0004.

Bernotat, Jasmin, Friederike Eyssel, and Janik Sachse. 2021. "The (Fe)Male Robot: How Robot Body Shape Impacts First Impressions and Trust Towards Robots." *International Journal of Social Robotics* 13 (3): 477–89. https://doi.org/10.1007/s12369-019-00562-7.

Blodgett, Su Lin, Solon Barocas, Hal Daumé III, and Hanna Wallach. 2020. "Language (Technology) Is Power: A Critical Survey of 'Bias' in NLP." arXiv. https://doi.org/10.48550/arXiv.2005.14050.

Briggs, R. A., and B. R. George. 2023. *What Even Is Gender?* London: Routledge. https://doi.org/10.4324/9781003053330

Brownlee, John. 2007. "Professor Ishiguro's Creepy Robot Doppelganger." *Wired*, April 26. https://www.wired.com/2007/04/professor-ishig/.

Bryson, Joanna J. 2018. "Patiency Is Not a Virtue: The Design of Intelligent Systems and Systems of Ethics." *Ethics and Information Technology* 20 (1): 15–26. https://doi.org/10.1007/s10676-018-9448-6.

Buddemeyer, Amanda, Erin Walker, and Malihe Alikhani. 2021. "Words of Wisdom: Representational Harms in Learning from AI Communication." arXiv. https://doi.org/10.48550/arXiv.2111.08581.

Buolamwini, Joy, and Timnit Gebru. 2018. "Gender Shades: Intersectional Accuracy Disparities in Commercial Gender Classification." *Proceedings of the 1st Conference on Fairness, Accountability and Transparency*, Proceedings of Machine Learning Research (PMLR) 81: 77–91.https://proceedings.mlr.press/v81/buolamwini18a.html.

Carpenter, Julie, Joan M. Davis, Norah Erwin-Stewart, Tiffany R. Lee, John D. Bransford, and Nancy Vye. 2009. "Gender Representation and Humanoid Robots Designed for Domestic Use." *International Journal of Social Robotics* 1 (3): 261–65. https://doi.org/10.1007/s12369-009-0016-4.

Cave, Stephen, and Kanta Dihal. 2020. "The Whiteness of AI." *Philosophy & Technology* 33 (4): 685–703. https://doi.org/10.1007/s13347-020-00415-6.

Coeckelbergh, Mark. 2020. *AI Ethics*. Cambridge, MA: MIT Press. https://mitpress.mit.edu/books/ai-ethics.

———. 2022. *Robot Ethics*. Cambridge, MA: MIT Press.

Cowls, Josh, Andreas Tsamados, Mariarosaria Taddeo, and Luciano Floridi. 2021. "A Definition, Benchmark and Database of AI for Social Good Initiatives." *Nature Machine Intelligence* 3 (2): 111–15. https://doi.org/10.1038/s42256-021-00296-0.

Dovidio, John F., Miles Hewstone, Peter Glick, and Victoria M. Esses. 2010. "Prejudice, Stereotyping and Discrimination: Theoretical and Empirical Overview." In *The SAGE Handbook of Prejudice, Stereotyping and Discrimination*, edited by John F. Dovidio, Miles Hewstone, Peter Glick, and Victoria M. Esses, 3–28. London: SAGE Publications Ltd. https://doi.org/10.4135/9781446200919.

Dufour, Florian, and Céline Ehrwein Nihan. 2016. "Do Robots Need to Be Stereotyped? Technical Characteristics as a Moderator of Gender Stereotyping." *Social Sciences* 5 (3): 27. https://doi.org/10.3390/socsci5030027.

Eyssel, Friederike, and Frank Hegel. 2012. "(S)He's Got the Look: Gender Stereotyping of Robots." *Journal of Applied Social Psychology* 42 (9): 2213–30. https://doi.org/10.1111/j.1559-1816.2012.00937.x.

Fazelpour, Sina, and David Danks. 2021. "Algorithmic Bias: Senses, Sources, Solutions." *Philosophy Compass* 16 (8): e12760. https://doi.org/10.1111/phc3.12760.

Floridi, Luciano, Josh Cowls, Thomas C. King, and Mariarosaria Taddeo. 2020. "How to Design AI for Social Good: Seven Essential Factors." *Science and Engineering Ethics* 26: 1771–96. https://doi.org/10.1007/s11948-020-00213-5.

Hundt, Andrew, William Agnew, Vicky Zeng, Severin Kacianka, and Matthew Gombolay. 2022. "Robots Enact Malignant Stereotypes." In *Proceedings of the 2022 ACM Conference on Fairness, Accountability, and Transparency*, 743–56. FAccT '22. New York, NY: Association for Computing Machinery. https://doi.org/10.1145/3531146.3533138.

Ihde, Don. 1990. *Technology and the Lifeworld: From Garden to Earth*. Bloomington: Indiana University Press.

Katzman, Jared, Angelina Wang, Morgan Scheuerman, Su Lin Blodgett, Kristen Laird, Hanna Wallach, and Solon Barocas. 2023. "Taxonomizing and Measuring Representational Harms: A Look at Image Tagging." arXiv. https://doi.org/10.48550/arXiv.2305.01776.

Lancaster, Karen. 2021. "Non-consensual Personified Sexbots: An Intrinsic Wrong." *Ethics and Information Technology* 23 (4): 589–600. https://doi.org/10.1007/s10676-021-09597-9.

Lee, Hee Rin, EunJeong Cheon, Maartje de Graaf, Patrícia Alves-Oliveira, Cristina Zaga, and James Young. 2019. "Robots for Social Good: Exploring Critical Design for HRI." In *2019 14th ACM/IEEE International Conference on Human–Robot Interaction (HRI)*, 681–82. https://doi.org/10.1109/HRI.2019.8673130.

Levy, David. 2008. *Love and Sex with Robots: The Evolution of Human–Robot Relationships*. New York: Harper Perennial.

Lupetti, M. L. 2022. "What Can We Do as Designers." In *Diversity Equity and Inclusion in Embodied AI: Reflecting on and Re-imagining Our Future with Embodied AI*, edited by Maria Luce Lupetti, Cristina Zaga, and Gijs Huisman. https://repository.tudelft.nl/islandora/object/uuid%3A17c24338-2850-40f7-b2bc-c8159f0d4836.

Malinowska, Joanna Karolina, and Tomasz Żuradzki. 2022. "Towards the Multileveled and Processual Conceptualisation of Racialised Individuals in Biomedical Research." *Synthese* 201 (1): 11. https://doi.org/10.1007/s11229-022-04004-2.

Mamak, Kamil. 2022. "Should Criminal Law Protect Love Relation with Robots?" *AI & SOCIETY*, April. https://doi.org/10.1007/s00146-022-01439-6.

———. 2023. *Robotics, AI and Criminal Law: Crimes Against Robots*. 1st edition. London: Routledge. https://doi.org/10.4324/9781003331100.

Mamak, Kamil, Pekka Mäkelä, and Raul Hakli. 2023. "Why Should We Pay Attention to Gender and Race in Robot Design?" *Technology and Sustainable Development 2023*, Conference Paper. Halden, Norway. https://techandsd.com/_files/mamak_etal_2023.pdf.

Mazzi, Francesca, and Luciano Floridi. 2023. *The Ethics of Artificial Intelligence for the Sustainable Development Goals*. Springer International Publishing.

Moore, Jared. 2019. "AI for Not Bad." *Frontiers in Big Data* 2. https://www.frontiersin.org/articles/10.3389/fdata.2019.00032.

Müller, Vincent C., and Michael Cannon. 2021. "Existential Risk from AI and Orthogonality: Can We Have It Both Ways?" *Ratio* 35 (1): 25–36. https://doi.org/10.1111/rati.12320.

Noble, Safiya Umoja. 2018. *Algorithms of Oppression: How Search Engines Reinforce Racism*. New York: New York University Press.

Nyholm, Sven, and Lily Eva Frank. 2019. "It Loves Me, It Loves Me Not: Is It Morally Problematic to Design Sex Robots that Appear to Love Their Owners?" *Techné: Research in Philosophy and Technology* 23 (3): 402–24. https://doi.org/10.5840/techne2019122110.

Perugia, Giulia, Stefano Guidi, Margherita Bicchi, and Oronzo Parlangeli. 2022. "The Shape of Our Bias: Perceived Age and Gender in the Humanoid Robots of the ABOT Database." In *Proceedings of the 2022 ACM/IEEE International Conference on Human–Robot Interaction*, 110–19. HRI '22. Sapporo, Hokkaido: IEEE Press.

Perugia, Giulia, and Dominika Lisy. 2023. "Robot's Gendering Trouble: A Scoping Review of Gendering Humanoid Robots and Its Effects on HRI." arXiv. https://doi.org/10.48550/arXiv.2207.01130.

Poulsen, Adam, Eduard Fosch-Villaronga, and Roger Andre Søraa. 2020. "Queering Machines." *Nature Machine Intelligence* 2 (3): 152. https://doi.org/10.1038/s42256-020-0157-6.

Riek, Laurel, and Don Howard. 2014. "A Code of Ethics for the Human–Robot Interaction Profession." SSRN Scholarly Paper. Rochester, NY. https://papers.ssrn.com/abstract=2757805.

Roesler, Eileen, Maris Heuring, and Linda Onnasch. 2023. "(Hu)Man-like Robots: The Impact of Anthropomorphism and Language on Perceived Robot Gender." *International Journal of Social Robotics*, March. https://doi.org/10.1007/s12369-023-00975-5.

Sætra, Henrik Skaug. 2021. "Loving Robots Changing Love: Towards a Practical Deficiency-Love." *Journal of Future Robot Life* (Preprint): 1–19. https://doi.org/10.3233/FRL-200023.

———. 2022. *AI for the Sustainable Development Goals*. 1st edition. CRC Press.

———, ed. 2023. *Technology and Sustainable Development: The Promise and Pitfalls of Techno-Solutionism*. New York: Routledge. https://doi.org/10.1201/9781003325086.

Sætra, Henrik Skaug, and John Danaher. 2023. "Resolving the Battle of Short- vs. Long-Term AI Risks." *AI and Ethics*, September. https://doi.org/10.1007/s43681-023-00336-y.

Shelby, Renee, Shalaleh Rismani, Kathryn Henne, AJung Moon, Negar Rostamzadeh, Paul Nicholas, N'Mah Yilla, Jess Gallegos, Andrew Smart, Emilio Garcia, and Gurleen Virk. 2023. "Sociotechnical Harms of Algorithmic Systems: Scoping a Taxonomy for Harm Reduction." arXiv. https://doi.org/10.48550/arXiv.2210.05791.

Shew, Ashley. 2020. "Ableism, Technoableism, and Future AI." *IEEE Technology and Society Magazine* 39 (1): 40–85. https://doi.org/10.1109/MTS.2020.2967492.

Spaccatini, Federica, Giulia Corlito, and Simona Sacchi. 2023. "New Dyads? The Effect of Social Robots' Anthropomorphization on Empathy towards Human Beings." *Computers in Human Behavior* 146 (September): 107821. https://doi.org/10.1016/j.chb.2023.107821.

Sparrow, Robert. 2020a. "Do Robots Have Race? Race, Social Construction, and HRI." *IEEE Robotics & Automation Magazine* 27 (3): 144–50. https://doi.org/10.1109/MRA.2019.2927372.

———. 2020b. "Robotics Has a Race Problem." *Science, Technology, & Human Values* 45 (3): 538–60. https://doi.org/10.1177/0162243919862862.

————. 2021. "How Robots Have Politics." In *The Oxford Handbook of Digital Ethics*, edited by Carissa Véliz, 292–311. https://doi.org/10.1093/oxfordhb/9780198857815.013.16.

Sparrow, Robert, Eliana Horn, and Friederike Eyssel. 2023. "Do Robots Have Sex? A Prolegomenon." *International Journal of Social Robotics* 15 (11): 1707–23. https://doi.org/10.1007/s12369-023-01052-7.

Stahl, Bernd Carsten, Doris Schroeder, and Rowena Rodrigues. 2023. "AI for Good and the SDGs." In *Ethics of Artificial Intelligence: Case Studies and Options for Addressing Ethical Challenges*, edited by Bernd Carsten Stahl, Doris Schroeder, and Rowena Rodrigues, 95–106. SpringerBriefs in Research and Innovation Governance. Cham: Springer International Publishing. https://doi.org/10.1007/978-3-031-17040-9_8.

Nature. 2023. "Stop Talking about Tomorrow's AI Doomsday When AI Poses Risks Today." *Nature* 618 (7967): 885–86. https://doi.org/10.1038/d41586-023-02094-7.

Strait, Megan, Ana Sánchez Ramos, Virginia Contreras, and Noemi Garcia. 2018. "Robots Racialized in the Likeness of Marginalized Social Identities Are Subject to Greater Dehumanization than Those Racialized as White." In *2018 27th IEEE International Symposium on Robot and Human Interactive Communication (RO-MAN)*, 452–57. https://doi.org/10.1109/ROMAN.2018.8525610.

Sullins, John P. 2012. "Robots, Love, and Sex: The Ethics of Building a Love Machine." *IEEE Transactions on Affective Computing* 3 (4): 398–409. https://doi.org/10.1109/T-AFFC.2012.31.

Tay, Benedict, Younbo Jung, and Taezoon Park. 2014. "When Stereotypes Meet Robots: The Double-edge Sword of Robot Gender and Personality in Human–Robot Interaction." *Computers in Human Behavior* 38 (September): 75–84. https://doi.org/10.1016/j.chb.2014.05.014.

Tay, Benedict Tiong Chee, Taezoon Park, Younbo Jung, Yeow Kee Tan, and Alvin Hong Yee Wong. 2013. "When Stereotypes Meet Robots: The Effect of Gender Stereotypes on People's Acceptance of a Security Robot." In *Engineering Psychology and Cognitive Ergonomics. Understanding Human Cognition*, edited by Don Harris, 261–70. Lecture Notes in Computer Science. Berlin, Heidelberg: Springer. https://doi.org/10.1007/978-3-642-39360-0_29.

Thomas, Veronica L., and Kendra Fowler. 2021. "Close Encounters of the AI Kind: Use of AI Influencers as Brand Endorsers." *Journal of Advertising* 50 (1): 11–25. https://doi.org/10.1080/00913367.2020.1810595.

Tomašev, Nenad, Julien Cornebise, Frank Hutter, Shakir Mohamed, Angela Picciariello, Bec Connelly, Danielle C. M. Belgrave, et al. 2020. "AI for Social Good: Unlocking the Opportunity for Positive Impact." *Nature Communications* 11 (1): 2468. https://doi.org/10.1038/s41467-020-15871-z.

Verbeek, Peter-Paul. 2011. *Moralizing Technology: Understanding and Designing the Morality of Things*. Chicago, IL: University of Chicago Press..

Vold, Karina, and Daniel R. Harris. 2021. "How Does Artificial Intelligence Pose an Existential Risk?" In *The Oxford Handbook of Digital Ethics*, edited by Carissa Véliz, 724–747. https://doi.org/10.1093/oxfordhb/9780198857815.013.36.

Wang, Angelina, Solon Barocas, Kristen Laird, and Hanna Wallach. 2022. "Measuring Representational Harms in Image Captioning." In *Proceedings of the 2022 ACM Conference on Fairness, Accountability, and Transparency*, 324–35. FAccT '22. New York, NY: Association for Computing Machinery. https://doi.org/10.1145/3531146.3533099.

Weßel, Merle, Niklas Ellerich-Groppe, and Mark Schweda. 2023. "Gender Stereotyping of Robotic Systems in Eldercare: An Exploratory Analysis of Ethical Problems and Possible Solutions." *International Journal of Social Robotics* 15 (11): 1963–76. https://doi.org/10.1007/s12369-021-00854-x.

Williams, Tom. 2023. "The Eye of the Robot Beholder: Ethical Risks of Representation, Recognition, and Reasoning over Identity Characteristics in Human–Robot Interaction". In *Companion of the 2023 ACM/IEEE International Conference on Human–Robot Interaction*, 1–10. HRI '23. https://doi.org/10.1145/3568294.3580031.

Winner, Langdon. 1980. "Do Artifacts Have Politics?" *Daedalus* 109 (1): 121–36.

Zaga, Cristina, M. L. Lupetti, and Gijs Huisman, eds. 2022. *Diversity Equity and Inclusion in Embodied AI: Reflecting on and Re-Imagining Our Future with Embodied AI*. https://doi.org/10.3990/1.9789036553599.

Złotowski, Jakub, Diane Proudfoot, Kumar Yogeeswaran, and Christoph Bartneck. 2015. "Anthropomorphism: Opportunities and Challenges in Human–Robot Interaction." *International Journal of Social Robotics* 7 (3): 347–60. https://doi.org/10.1007/s12369-014-0267-6.

4 | Human-like Ethics and Responsibility of Robots

Introduction

When discussing human-like robots, one might think that their resemblance to humans will not be limited to external shells, but also extend to their behaviors. That human-like robot will be almost the same, except for the matters from which both kinds of entities are built. To what extent are these scenarios possible? The previous two chapters focused on the external aspects of human likeness that do not pose a serious challenge to contemporary technology. This chapter is different, and it deals with one of the main concerns about the ethical issues related to artificial intelligence (AI) and robots, namely, the ethical behavior of AI/robots and responsibility for actions involving such tools (cf. e.g., Dignum 2019; Kokkonen 2020). This chapter focuses on those issues from the perspective of the main frame of this book, which is human likeness. There will be engagement with two questions that involve that frame. First, can robots have human-like ethics? Second, can robots be responsible in a human-like sense? These issues, while being only a subsection of the general issues of robot ethics and the responsibility for robot actions, are still issues that cover many problems that cannot be resolved fully in the length of a single chapter. Instead, the aim of this chapter is to show why the task of creating robots that have human-like ethics and are responsible in the human sense is harder than it might seem to be, or even impossible. Before going into those questions, let us consider scenarios that will help to illustrate the matter of the chapter.

In the first scenario, a robotic company was asked about deploying their promising humanoid robots as help at international airports. The company that commissioned the offer underlined that the robots will work in adverse conditions as the first line of contact with travelers from all around the world who are in a hurry, are confused, or have missed a flight. They underlined that robots will be like employees of the airport, and that the airport management wanted the robots to make a good impression on the travelers, which will require ethical sensitivity. In the beginning, the company set up a team to implement ethical rules in their robots. After a couple of months, they thought that their robots were ready for testing in the wild. The test robot was deployed in a small airport. A team of evaluators was constantly monitoring every interaction of the robot. Additionally, the other members of the team interviewed the travelers, obtaining their feedback after their interactions with the constantly monitored robot. At the end of the day, the team was devastated. The complexity of the interactions overwhelmed them. Naming a few problems, they said that they did not anticipate that the travelers would flirt with robots; the robot in one of those interactions behaved rudely by saying that the shirt of the person was dirty from sauce. It was a response to the question, "How do you like me." The answer was true; there was, in fact, a dirty patch on the shirt. However, the person reported being truly offended by that comment. Another traveler complained that the robot ignored him after initially their eyes met and the robot nodded its head, which the traveler understood as readiness to help. Instead, without a word of explanation, the robot chose to interact with a family of four that announced their need for help at more or less the same time. There was also a group of people from foreign countries who noted that robots seemed to ignore their attempts at communication that were made in the same way in which help is usually asked for in their home country. Humans would guess the intention while looking at the signals, even if those signals were being seen for the first time. Members of the team went back to work. The next attempt was more successful, but because there were always some new problems, for example, words acquiring new meanings and becoming more hostile, or norms changing over time, it was decided to keep the team that was responsible for ethics as a permanent component of their service with robots.

In another scenario, a shifty person specializing in stealing got into trouble with the law. To avoid the direct risk of exposure, they decided to buy a child-like humanoid robot to continue their criminal activities indirectly. The robot's cover was to be a regular shopper; it justified having a shopping bag for items and its presence in different shops. The owner modified his robot to steal relatively small, cheap items. A single act is not that harmful to the victims to the extent that they barely notice the missing thing, or even if they do, they do not bother to inform law enforcement agencies about their loss. The robot was also prepared for several options. When it was caught red-handed, keeping the stolen item in hand, it was made to learn to gently apologize, nod, and promise that it would never do it again, which was accompanied by a facial expression that expressed sincere embarrassment. The owner of the robot learned that it usually ended the case, and the owner of the shop usually granted forgiveness upon seeing the embarrassed face. After long months of successful operations of the robot without serious problems, which led him to earn a considerable sum of money, the owner became more careless. The robot was too often present in the same marketplaces. The owner abandoned his initial rule that the robot would operate in one place and then would switch locations to avoid suspicion. The owner's problems started when a security guard, out of curiosity, started to watch a cute-looking robot who had been shopping in the same marketplace many times. Out of boredom, he followed the robot on cameras and shockingly observed that besides regular shopping, it was collecting extra items without paying for them. The situation was repeated a few times, and the security guard informed the authorities that followed the robot and unearthed the organized nature of the criminal activities.

In the final scenario, a robotic company was looking for an opportunity for growth and decided to enter the healthcare business. An aging society and the growing need for medical care seemed to be strong arguments for investing time and resources in that business. They looked for their niche and decided to specialize in the risky field, offering humanoid robot nurses that dispense and administer drugs. They were hoping to be financed by medical insurance. They worked on their product, and after testing almost countless options, they thought they were ready to enter the market. After successful trials and positive reactions, they decided to extend their offer, claiming that the robot nurse could dispense drugs to any patient. Unfortunately, one of the robots dispensed a lethal dose of a drug. The news immediately reached the media, and the CEO of the company nervously published a video on his social media records with his smartphone, blaming individual robots for mistakes and trying to discharge their liability. After a few minutes, seeing the outrage, they deleted the initial video and published a statement in which a detailed investigation was promised. During the long procedure, lawyers representing the robotic company tried to convince the investigator that the mistake could not be attributed to anyone in the company, that the robot learned how to behave via interactions with patients, and there was a probability that the unfortunate patient will incorrectly use a nearly perfect mechanism. They proposed destroying the robot that was involved in the accident.

This chapter is organized as follows. After the introduction, there are two sections that relate to the main issues. First, there is a discussion about the safety issues related to the human-like ethics of robots. Second, there is a discussion on whether robots could be responsible in a human-like sense. The chapter ends with conclusions.

Human-like Ethics of Robots

In this section, I discuss the safety issues of robots, which is one of the main worries related to their wider deployment. Machines that have bodies that can interact with external environments, including humans and their fragile bodies, raise justifiable worries. This particularly applies to humanoid robots that are meant to be around humans and not, for example, in the controlled environment of a factory in which access is limited. Those safety worries are strengthened by pop culture in which robots are often presented as entities that threaten human existence.

The safety of humanoid robots covers a range of issues, such as making sure that the materials from which they are built are not toxic. The other understanding of safety is related to the human-like form of robots, which is the subject of Chapter 2. There is a discussion on the risks that are related to populating the world with human-like entities. The risks described there are about human behaviors toward other humans but related to the robots that are around. In this chapter, safety is related to how robots behave. The direct source of risk is in robots. These issues are usually a matter of concern for machine/robot ethics (see Chapter 1). Nath and Sahu point out that machine ethics is different from other ethics of technology. They give an example of computer ethics that is focused on the ethics of humans' use of computers. They underline that the task of machine ethics is to ensure that the machines' behaviors are ethical (Nath and Sahu 2020: 103). In general, it is assumed that robots should adopt human-like ethics to be safe. Some are optimistic that robots could be ethical in the human sense, and even surpass humans in that aspect of life. For example, in a book on robot ethics and the innovation economy, Johannessen wrote: "Intelligent robots will be increasingly able to reflect on ethics and will be more able to act ethically than humans" (Johannessen 2021: 1). In the following, the focus is on skeptical views on that topic. The plan is to focus only on those positions that challenge the question of human-level ethics in robots. There will be an overview of the various positions that raise doubts that robots can have human-like ethics.

There are three basic approaches to implementing ethics in machines: top-down, bottom-up, and hybrid (cf. Allen et al. 2005). In short, according to Allen et al., top-down starts with ethical rules that machines need to follow. It draws on some ethical theories, like utilitarianism or deontology, that should be applied by machines when morally problematic events occur. The main weakness of this approach is that it is not well-equipped for the complexity of what can happen in the wild when the machine is deployed. The bottom-up approach focuses on learning ethics from observation and experience. It starts with real-world examples and end scenarios. One of the problems of this approach is concern with the learning material and biases in data that could lead to unethical outcomes. The hybrid approach takes advantage of both of them, starting with predefined rules, and allowing learning from examples (Allen et al. 2005).

These approaches are more problematic when it comes to details. As mentioned, in top-down approaches, the starting point is the ethical theory to be applied. The main theories that are usually mentioned in that context are deontology and utilitarianism. Let us illustrate the problems that might arise from applying those theories to robots. A good example with deontological theory is the issue of deception in robotics. Yew points out that if we look at the problem of deception from the perspective of deontology, the default position should be a claim that the deception of robots is morally wrong, and offends the virtues of honesty and trustworthiness (Yew 2021). However, it has been pointed out that deception is not only acceptable but also necessary in fields like the military, healthcare, and education (cf. Shim and Arkin 2013; Danaher 2020). Isaac and Bridewell argue that robots should be able to use "white lies" when communicating. Without those deceptive practices, communication with robots cannot be natural (Isaac and Bridewell 2017). The robot from the second scenario told the truth about a traveler's outfit, but the context of the conversation would justify saying that the person looks good.

Utilitarian theory could also be flawed in practice. For example, while evaluating a robot's action, whether the action is adjudged to be good or bad from a utilitarian point of view, there is a need to decide what to count as morally important aspects of the consequences of the action. Recently, Singer and Tse called for broader inclusion of the interests of animals within AI ethics, which, in their opinion, is rarely mentioned in the field (Singer and Tse 2023; see also Hagendorff et al. 2022). So, according to their call, in an ethical assessment of actions, they should consider not only the interests of humans, but also those of animals. One of the hypothetical examples is Bostrom's "paperclip maximizer." (Bostrom 2003). In this scenario, the AI is given the goal of making paperclips, and having that goal, AI is using all of Earth's resources to manufacture paperclips, thereby putting humanity in danger. It is an example usually used in the context of discussing existential risks related to the emergence of artificial general intelligence (AGI). (For more on AGI and existential risks, see Vold and Harris 2021). This example is exaggerated and does not do justice to utilitarian theory, but the reason to refer to it is to show issues with the simple application of utilitarian theory without thinking about all the nuances that could matter in practice.

These dominant theories—utilitarianism and deontology—have their own problems, but it has also been pointed out that focusing on them marginalizes other approaches, including virtue ethics, as well as non-Western traditions like Ubuntu, Confucian, Shinto, and Buddhist ethics, which are less represented in literature (cf. Segun 2021). In their paper, Génova et al. explored the issue of omission of virtue ethics, which, according to the authors, is more akin to non-Western tradition in the discussion on robot ethics. The authors believe that this is due to the theoretical problem of transferring the ethical framework into rules that could be adopted. Measuring utility seems to be more operationalized in the context of technology than the call for being a good person (Génova et al. 2023: 5–6). Génova et al. believe that ethics of robots could not be limited to teaching robots merely to follow a moral code because being good is something more complex (Génova et al. 2023). According to them, ethics in the context of robots is not just a technical problem that can be solved with better technology, and human ethics encompass the ability to solve problems and the capacity to identify the problems that are worth solving (Génova et al. 2023: 2). They asked a final question in their paper, which is also interesting from the perspective of the general framework of this chapter: "Can machines learn ethics like humans do?" and they answer that question in the negative. In a similar vein, Hutler et al. argue that ethics for robots cannot be reduced to programming human morality into robots. Human morality needs to be rethought and retheorized to fit robots. They underline that ethics need to be adapted to real robots with their technological limitations that are already present or could be developed in the near future, and not to robots that share human-like qualities, like general intelligence (Hutler et al. 2023: 2). Also, Sparrow shares a pessimistic view on the topic of robots having human-like ethics due to the complexity of human ethics. In the paper titled "Why Machines Cannot Be Moral," drawing on the works of

Raymond Gaita, Sparrow argues that ethical dilemmas are problems for real people and not problems for everyone. What is more, the important element in evaluating ethical claims is the personal history of the person facing an ethical problem. This allows Sparrow to formulate the claim that the whole project of building ethics into machines is flawed and, at best, could provide a shallow semblance of ethics (Sparrow 2021).

Brożek and Janik are also skeptical about the project of creating human-like morality in robots. They point out that, ironically, robots could be better than humans in being an approximation of idealized Kantian or utilitarian ethics, but they add that being moral in a human way is something more complex (Brożek and Janik 2019). They explain that there are external and internal conditions of moral agency. The external, in brief, is related to recognition as such by the members of the community. That element, according to them, could be met. More problematic is the internal element which, in their view, is fundamentally related to emotions that robots lack. They add that emotions in machines cannot be limited to the recognition of emotions in humans but rather should be "the driving force behind the machine's behavior" (Brożek and Janik 2019: 101). Coeckelbergh notes that the standard necessary conditions for having emotions are consciousness, mental states, and feelings, which these current robots do not have (Coeckelbergh 2010).

Another problematic aspect of implementing ethics in robots is the norms that robots should follow to benefit users and society. Coggins and Steinert discuss widely these issues and argue that there are seven problems with norm-complaint robots, which are: norm biases, paternalism, tyrannies of majority, pluralistic ignorance, paths of least resistance, outdated norms, and technological-induced norm change (Coggins and Steinert 2023). They argue that even if it is possible for robots to comply with social norms, it could still lead to socially questionable outcomes. Now, I will briefly sum up these problems. Norm biases refer to the problem that different groups could have different social norms and choosing a robot to comply with norms that are accepted in one group could lead to the transgression of others. The authors use the example of different ways of greeting in different cultures; the robot that welcomes people in one way will ignore the norms that are accepted in cultures that adopted alternative ways of welcoming. In this context, it is worth mentioning the Moral Machine experiment, which collected data from people all around the world about ethical preferences in scenarios involving autonomous cars. People were asked to reveal preferences in dilemmas on the road in a situation in which an accident was unavoidable. This experiment showed the varieties of preferences, including cross-cultural ethical variations (Awad et al. 2018).

The second problem, called "paternalism," refers to the suggestion that experts should first choose which norms robots should adopt. They use the example of a robot that refuses to respond if the command of the user contains swearing. Those who swear would need to comply with the norms that someone else adopted. The "tyranny of the majority" is a problem that is partially related to paternalism. In this case, the decision about which norm is valid was not made by a preselected group of people but represents the expectation of the majority. In a democracy, as the authors note, the groups that do not have representatives who could impact them could experience harm from the decisions that are accepted by the majority. The more a given society is polarized, the less unacceptable the decisions of the majority might be to the minorities.

"Pluralistic ignorance" refers to social phenomenon in which members of the groups could be reluctant to reveal their real opinion on some topics due to the fear that other members of the group would disagree. As a result, people could publicly endorse some norms that they dislike in private (on the concept of "pluristic ignorance," see, .e.g., J. O'Gorman 1986; Bjerring et al. 2014; Halbesleben and Buckley 2004). Fifth, "paths of least resistance" refers to the tendency to follow norms even if we do not endorse them anymore because it is easier to repeat the past than deviate from them. They use an example of gender discrimination in a work context. Even if members of the groups in some companies believe that women and men should have equal opportunities and that archaic norms should not be adopted, the change of behavior takes time. The sixth problem focuses on the fact that norms could be outdated, and whether robots could keep up with them. Finally, seventh, which is related to the previous one, the change in moral norms could be related to the robots themselves (on moral change in the context of technology, see, e.g., Danaher and Sætra 2022, 2023; Hopster et al. 2022). Coggins and Steinert conclude in their paper that we cannot uncritically rely on norm-compliant robots for the aforementioned reasons. Acting according to norms, they do not call for the abandonment of work on building norm-compliant robots, but rather to treat the agenda about the whole endeavor more critically.

Wallach and Allen reserve a human-like level of morality for humans but argue that morality is a spectrum, and robots could have been situated on that spectrum. They note that at the low end of sensitivity to morally relevant facts is "operational morality." The moral significance lies only in the hands of humans—designers and users. The more sophisticated machines could have "functional morality," in which the focus is on the machines themselves, which could have the capacity for assessing and responding to moral challenges. "Functional morality" is between "operational morality" and full moral agency that (some) humans have. "Functional morality" comes in degrees. It could concern machines with significant autonomy levels and relatively low sensitivity to values, and vice versa, with little autonomy and high sensitivity to values (Wallach and Allen 2009: 25–26).

As was shown in that selected overview, the task of creating human-like ethics meets with complicated challenges. There are many issues as to why human-like ethics in robots is dubious. Some believe that ethics in the human sense is almost reserved for humans. It was particularly mentioned that robots lack emotions, which some see as necessary elements of human ethical life. Issues were raised with ethical norms that robots would need to follow, which might be biased or outdated. In general, the common core of the critical positions is the notion that ethics is a complex phenomenon that cannot be limited to programmable rules. It is complex not in the sense that it poses a challenge to computational power, but rather in the sense of not being a matter that is programmable. What seems to be more possible to achieve, in the light of the presented positions, is a less ambitious agenda. A kind of ethics that is similar to human-like ethics to some extent, but it is not in the full sense. However, having not-fully-human-like ethics in machines that are around us entails serious consequences.

It can be said that robots will not be like children, who learn in their early years how to be ethical, and then we might forget about that process. In the case of robots, this learning phase, if we could use that analogy, would be constant, and we, as a society, need to control whether our values are preserved in robots' actions permanently. There will be a constant burden on humans to make sure that robots are behaving in a way that, in the case of humans, would be ethical. It will be no job that needs to be done once, and then we could benefit from its effects. It will be permanent work that requires monitoring, calibrating, changing, and rethinking ethical guidelines for robots. More accurate here could be an analogy to wild, dangerous animals, which, even if domesticated and trained and behaving properly for a long time, can never be trusted truly. In other words, robots could have some ethics, but not in the sense that fully corresponds with the human level.

Human-like Responsibility

The second main issue that is covered in this chapter is related to the problem of responsibility, which is different from the matter of having ethics. As Nyholm points out, working upon morally expected principles and being responsible for actions are different issues (Nyholm 2020: 55). This part of the chapter is focused on the second issue, which is captured in the question, "Can robots be responsible in a human-like sense?" This section provides an overview of the selected positions that challenge the issue of it being possible for robots to be responsible in a human-like sense. This problem required more introductory work than the previous one. The term responsibility is ambiguous; different authors focus on different aspects of that concept while formulating positions regarding robots. The main message in this section is that robots cannot be *fully* responsible in a human-like sense in all the senses of responsibility. Before I focus on problematizing the answer to that question, I need to use a few words of introduction on why this is a problem in the first place, and then present the different views on responsibility.

Let us turn to the issue of how the deployment of AI-based tools is associated with issues on the grounds of responsibility. In the seminal paper on this topic, Matthias points out that the deployment of the "learning automata," as he called it, poses a challenge in ascribing moral and legal responsibility for outcomes of the operation of new technologies (Matthias 2004). A few years later, Sparrow, in a paper titled "Killer Robots," touched on a similar problem describing the potential use of robots on the battlefield and the issues with ascribing responsibility (Sparrow 2007). In its classical meaning, following Rudy-Hiller, to be responsible requires demonstrating that a person (agent) passes jointly two conditions to be blamed or praised for action: control condition (Was the person *free* in performing action?) and epistemic (also: knowledge, cognitive, mental) condition (Was the person aware of what they are doing?) (see Rudy-Hiller 2018; see also Coeckelbergh 2020: 109–111). The deployment of new technologies, and especially AI-enabled tools, challenges those conditions. Matthias noted that a system that has some autonomy and can learn from the environment and adapt to new circumstances, is problematic for classical requirements of responsibility (Matthias 2004). The harms resulting from the malfunction of technologies could not be applicable to humans (designers or users), who could not be held responsible for actions performed by AI tools. If we do not resign from using such tools, which Matthias predicted to not be a realistic scenario, we need to accept the emergence of "responsibility gaps" (but, see Tigard 2021).

Gunkel described three responses to responsibility gaps: first, always ascribe responsibility to humans, regardless of the issue of how advanced the machine is. The second response is shared responsibility between humans and machines (see also Nyholm 2018); third, it introduces the quasi-responsibility of machines (Gunkel 2020). Ascribing responsibility to humans for the actions of robots would require abandoning the classical understanding of responsibility that was mentioned earlier. The two remaining responses—shared responsibly and the responsibility of robots—are also problematic. For example, Hakli and Mäkelä argue that the responsibility of machines, as well as shared responsibility, is excluded. They explain that the responsible could be moral agents and shared responsibility could only happen between moral agents, and thus responsibility could not be shared with entities that could not be responsible for themselves. According to them, robots will never be able to be responsible in a human-like sense (Hakli and Mäkelä 2019). Gogoshin, summarizing the state of the discussion on that matter, wrote, "It is almost a foregone conclusion that robots cannot be morally responsible agents" (Gogoshin 2021: 1). If robots could not be moral agents, then not only moral but also criminal responsibility could be excluded. Asaro underlines that criminal action requires a moral agent who could be held responsible (Asaro 2012: 181). Robots that cannot have moral agency cannot be responsible. Asaro sums up his deliberations in the context of criminal responsibility and robots with a statement concerning robots that they have: "A body to kick, but still no soul to damn" (Asaro 2012: 169).

This short overview shows that, at least for now, from the perspective of the classical understanding of responsibility, there is an impasse. The deployment of robots operated with AI tools might cause problems for which no one could be held responsible in the classical sense when harm occurs. Why would it be problematic to accept the emergence of responsibility gaps? Danaher, for example, points out that the responsibility gap could lead to a retribution gap. He explains that people have a natural tendency to look for someone to punish when harm happens, and if there is no one to punish, it might lead to scapegoating at an individual level, at a more abstract level, to undermine the trust in the rule of law (Danaher 2016).

The claim that contemporary robots could not be responsible for their actions is a dominant view, though the claim that robots are not able to be responsible is not universally accepted and some challenge this position. There are views that accept the possibility of robots being subject to responsibility, at least theoretically, including being subject to punishments (see, e.g., Hallevy 2013; Lagioia and Sartor 2020; Simmler and Markwalder 2019). However, even if one believes that responsibility might be extended to robots, it will not be the exact same responsibility as it is in the case of humans. Robots would need to adapt the structures and concepts related to responsibility. For example, Lagioia and Sartor point out that "a convenient system of penalties and due adjustments has to be designed" (Lagioia and Sartor 2020: 460). So even if it were possible, it still would not be responsibility in a human-like sense. Elsewhere, I point out that the laws that cover human behaviors are made for people with human biology (Mamak 2022). The differences in the way in which entities are built would require having other systems of rules that would govern the behaviors, including rules ascribing responsibility, including punishments.

While discussing the possible responsibility of robots, some think about the threshold for human-like responsibility that robots could someday cross (Pagallo 2013: 38–40; Hu 2018: 487). While theoretically this option could not be excluded, there is a problem of establishing the set of qualities for being responsible in a human sense. There is also an epistemological problem: how would we know that robots do have those qualities? In his recent book, Gunkel points out the issue of qualities of human likeness, as well as epistemological issues. In his opinion, these two issues are almost equally problematic. He discusses these issues in the context of the question of whether robots could be "natural persons" in the book that that tries to situate robots in the moral and legal realms (see Chapter 4 in Gunkel 2023).In the next chapter of this book, there is also a discussion on that topic in the context of the question of robots being moral patients (see Chapter 5).

A different take on the problem of responsibility is not to focus on the robot's ontology, in the sense of entities that might be held responsible, but instead on their performance in morally relevant contexts. For example, Sullins argues that it might be sufficient in some circumstances not to be a moral agent but to be seen as such. He gives three requirements for robots to be seen as moral agents. First is significant autonomy from human creators or operators. Second, they ascribe to their intentions. Third, showing an understanding of responsibility (Sullins 2006). This lowers the bar for robots to be actors in moral practices.

A similar line of thought is represented by Coeckelbergh, who, in one of their earlier works, uses the concept of virtual moral responsibility. Instead of focusing on the "inside" of the agent that could be responsible,

Coeckelbergh focuses on the appearances. He argues that what matters in that respect is not the qualities of the entity but its performance (Coeckelbergh 2009; about the appearance of morality, see also Coeckelbergh 2010). More recently, Gogoshin, drawing on Strawsonian views on responsibility (Strawson 1962), argues that robots could be morally responsible if they comply with the normative expectations of the moral community (Gogoshin 2021; see also other Strawsonian approach, Tigard 2021). She underlines the community's role in deciding who its members are. It is not focused that much on the subject of responsibility but on the moral practices of communities.

Would robots be accepted or seen as responsible agents by humans? Below are some of the experiments that have focused on different aspects of responsibility practices and robots from the perspective of human–robot interactions. These experiments suggest that, at least partially, the participants are attributing some responsibility to robots, and could include them in moral practices. In the experiment conducted by Kahn et al., the robot unfairly deprived the participant of the experiment cash reward (20 USD), and 65% of the participants attributed accountability to the robot for that (cf. Kahn et al. 2012). Experiments by Lee et al. focus on strategies for dealing with robot breakdowns. When such a thing happened, to mitigate the negative impression of an error, robots were asked to apologize for it, and some participants accepted these apologies (Lee et al. 2010). More recently, Lee et al. noticed in their experiments that people might want to punish robots (Lee et al. 2021). In their experiment, Stuart and Kneer presented participants with a situation in which different agents knowingly ran the risk of bringing about substantial harm; those agents were human, group agents, and AI-driven robots. They found, among other things, that judgments of wrongness and blame were relatively similar across these types of agents, which also include robots (Stuart and Kneer 2021). In the study examining perceptions on the punishment of robots, Lima et al. found "that even though participants are aware that robots and AI do not satisfy these preconditions, they, nonetheless, believe electronic agents should be punished" (Lima et al. 2020: 4).

The aforementioned responses to robots' wrongs, while suggesting that robots could be included in moral practices, at least partially, could be seen as signs of possible problems. The experiment conducted by Feier et al. shows that humans who delegate tasks that lead to bad outcomes for machines may be judged with more leniency than those humans who delegate tasks to other humans. As the authors show, this result has troubling implications; for example, companies might want to use that effect to shift responsibility and hide behind algorithms. They add that this is an additional argument for regulating the issue of responsibility to prevent the possibility of using that effect (Feier et al. 2022). In other words, this study shows that some bad actors might want to use robots to escape from responsibility by using humans' willingness to blame robots. Coeckelbergh notes that "even if robots counted as responsible agents in some sense, that does not necessarily diminish the responsibility of other responsible agents, including human responsible agents" (Coeckelbergh 2022: 131).

Trying to summarize the views that accept that it is for the community to decide to include robots in moral practice, if I have interpreted those positions correctly, they are meant to be applicable only to some extent, not to cover all morally relevant actions of robots. Starting with Sullins (2006), he presents his view as not being universally applicable, which could cover all situations in which robots are involved, but instead he underlines that in some cases, it could be enough to treat robots as responsible agents. Gogoshin (2021), in her position, refers to the moral community expectations, which could differ if the gravity of consequences was different (e.g., causing death could be treated differently from hitting someone lightly on a narrow street while trying to overtake). Also, Coeckelbergh, while underlining the possibility of making robots responsible, does not exclude making humans responsible for some consequences.

We return to this section's main question, "Can robots be responsible in a human-like sense?" According to the presented positions, it can be said that robots could never be fully responsible like humans. Some believe that robots are not, or even never could be, morally responsible agents in the traditional understanding of responsibility, which is the dominant position. Under that view, the question on the main question is the most straightforward: No; more challenging responses are required in the remaining views.

Some argue that robots could cross the threshold of responsibility now or at some point in the future. Those who accept the proposition of responsibility note the need to adjust their response to robots. It does not mean that robots will be responsible in the exact same sense, for example, being jailed for their wrongs. The responsibility would need to be adopted by robots, which are built in ways different from humans.

There are also positions that suggest shifting the focus of the interests; instead of discussing the features of the robots as potential holders of responsibility, we should observe how members of the moral community respond to the morally relevant actions of robots. If robots could be seen as morally responsible agents, then it might be enough to consider them as members of that moral club. But also, it seems that the applicability is limited and could be dependent on the gravity of the consequences; in some cases, involving humans in responsibility practices might be required.

Conclusions

This chapter dealt with two main questions. First, can robots have human-like ethics; second, can robots be responsible in a human-like sense? A positive answer to these two questions seems dubious, if not impossible. As far as human-like ethics is concerned, this chapter refers to positions that problematize the task of implementing human-like ethics into robots. Ethics, in those views, seems to be a complex issue that cannot be easily programmable. Some underline the meaning of human qualities for the ability to be ethical in a human sense, such as emotions, that robots, at least for now, cannot have. Some underline that this task seems impossible to achieve due to the varieties of approaches to ethics around the world. The project of implementing ethics into robots needs to be less ambitious than achieving human-like level. However, having robots with non-human-like ethics would entail a burden on humans and a need for constant control over the ethical life of robots.

The matter of responsibility in a human-like sense in robots is no less problematic. The dominant view is that robots cannot now be responsible in a human-like sense. Some are going even further, arguing that robots would never be as responsible as humans are. Skeptical positions are formulated upon a classical understanding of moral responsibility, which requires moral agents to whom responsibility could be attributed. However, there are positions that allow for robots to be responsible and even punished, though in a way that is suitable for them, not in the same ways that humans are. There are views that link the problem of responsibility of robots with moral practices of human communities. If robots could be seen as responsible agents, people would attribute human-like responsibility to them or apply practices related to responsibility to them; then it might be enough to accept robots' responsibility, at least partially. However, the applicability of the last group of views seems to be limited, and there could be cases in which the responsibility of human beings would need to be included.

References

Allen, Colin, Iva Smit, and Wendell Wallach. 2005. "Artificial Morality: Top-down, Bottom-up, and Hybrid Approaches." *Ethics and Information Technology* 7 (3): 149–55. https://doi.org/10.1007/s10676-006-0004-4.

Asaro, Peter M. 2012. "A Body to Kick, but Still No Soul to Damn: Legal Perspectives on Robotics." In *Robot Ethics: The Ethical and Social Implications of Robotics*, edited by Patrick Lin, Keith Abney, and George A. Bekey. Cambridge, MA: MIT Press.

Awad, Edmond, Sohan Dsouza, Richard Kim, Jonathan Schulz, Joseph Henrich, Azim Shariff, Jean-François Bonnefon, and Iyad Rahwan. 2018. "The Moral Machine Experiment." *Nature* 563 (7729): 59–64. https://doi.org/10.1038/s41586-018-0637-6.

Bjerring, Jens Christian, Jens Ulrik Hansen, and Nikolaj Jang Lee Linding Pedersen. 2014. "On the Rationality of Pluralistic Ignorance." *Synthese* 191 (11): 2445–70. https://doi.org/10.1007/s11229-014-0434-1.

Bostrom, Nick. 2003. "Are You Living in a Computer Simulation?" *Philosophical Quarterly* 53 (211): 243–55.

Brożek, Bartosz, and Bartosz Janik. 2019. "Can Artificial Intelligences Be Moral Agents?" *New Ideas in Psychology* 54 (August): 101–106. https://doi.org/10.1016/j.newideapsych.2018.12.002.

Coeckelbergh, Mark. 2009. "Virtual Moral Agency, Virtual Moral Responsibility: On the Moral Significance of the Appearance, Perception, and Performance of Artificial Agents." *AI & SOCIETY* 24 (2): 181–89. https://doi.org/10.1007/s00146-009-0208-3.

———. 2010. "Moral Appearances: Emotions, Robots, and Human Morality." *Ethics and Information Technology* 12 (3): 235–41. https://doi.org/10.1007/s10676-010-9221-y.

———. 2020. *AI Ethics*. Cambridge, MA: MIT Press. https://mitpress.mit.edu/books/ai-ethics.

———. 2022. *Robot Ethics*. Cambridge, MA: MIT Press.

Coggins, Tom N., and Steffen Steinert. 2023. "The Seven Troubles with Norm-compliant Robots." *Ethics and Information Technology* 25 (2): 29. https://doi.org/10.1007/s10676-023-09701-1.

Danaher, John. 2016. "Robots, Law and the Retribution Gap." *Ethics and Information Technology* 18 (4): 299–309. https://doi.org/10.1007/s10676-016-9403-3.

———. 2020. "Robot Betrayal: A Guide to the Ethics of Robotic Deception." *Ethics and Information Technology* 22 (2): 117–28. https://doi.org/10.1007/s10676-019-09520-3.

Danaher, John, and Henrik Skaug Sætra. 2022. "Technology and Moral Change: The Transformation of Truth and Trust." *Ethics and Information Technology* 24 (3): 35. https://doi.org/10.1007/s10676-022-09661-y.

———. 2023. "Mechanisms of Techno-Moral Change: A Taxonomy and Overview." *Ethical Theory and Moral Practice*, June. https://doi.org/10.1007/s10677-023-10397-x.

Dignum, Virginia. 2019. *Responsible Artificial Intelligence: How to Develop and Use AI in a Responsible Way.* Artificial Intelligence: Foundations, Theory, and Algorithms. Cham: Springer International Publishing. https://doi.org/10.1007/978-3-030-30371-6.

Feier, Till, Jan Gogoll, and Matthias Uhl. 2022. "Hiding Behind Machines: Artificial Agents May Help to Evade Punishment." *Science and Engineering Ethics* 28 (2): 19. https://doi.org/10.1007/s11948-022-00372-7.

Génova, Gonzalo, Valentín Moreno, and M. Rosario González. 2023. "Machine Ethics: Do Androids Dream of Being Good People?" *Science and Engineering Ethics* 29 (2): 10. https://doi.org/10.1007/s11948-023-00433-5.

Gogoshin, Dane Leigh. 2021. "Robot Responsibility and Moral Community." *Frontiers in Robotics and AI* 8: 342. https://doi.org/10.3389/frobt.2021.768092.

Gunkel, David J. 2020. "Mind the Gap: Responsible Robotics and the Problem of Responsibility." *Ethics and Information Technology* 22 (4): 307–20. https://doi.org/10.1007/s10676-017-9428-2.

———. 2023. *Person, Thing, Robot: A Moral and Legal Ontology for the 21st Century and Beyond.* Cambridge, MA: MIT Press.

Hagendorff, Thilo, Leonie Bossert, Tse Yip Fai, and Peter Singer. 2022. "Speciesist Bias in AI—How AI Applications Perpetuate Discrimination and Unfair Outcomes against Animals." arXiv. https://doi.org/10.48550/arXiv.2202.10848.

Hakli, Raul, and Pekka Mäkelä. 2019. "Moral Responsibility of Robots and Hybrid Agents." *The Monist* 102 (2): 259–75. https://doi.org/10.1093/monist/onz009.

Halbesleben, Jonathon R. B., and M. Ronald Buckley. 2004. "Pluralistic Ignorance: Historical Development and Organizational Applications." *Management Decision* 42 (1): 126–38. https://doi.org/10.1108/00251740310495081.

Hallevy, Gabriel. 2013. *When Robots Kill: Artificial Intelligence under Criminal Law.* Boston: Northeastern.

Hopster, J. K. G., C. Arora, C. Blunden, C. Eriksen, L. E. Frank, J. S. Hermann, M. B. O. T. Klenk, E. R. H. O'Neill, and S. Steinert. 2022. "Pistols, Pills, Pork and Ploughs: The Structure of Technomoral Revolutions." *Inquiry*, 1–33. https://doi.org/10.1080/0020174X.2022.2090434.

Hu, Ying. 2018. "Robot Criminals." *University of Michigan Journal of Law Reform* 52: 487.

Hutler, Brian, Travis N. Rieder, Debra J. H. Mathews, David A. Handelman, and Ariel M. Greenberg. 2023. "Designing Robots That Do No Harm: Understanding the Challenges of Ethics for Robots." *AI and Ethics*, April. https://doi.org/10.1007/s43681-023-00283-8.

Isaac, Alistair, and Will Bridewell. 2017. "White Lies on Silver Tongues: Why Robots Need to Deceive (and How)." In *Robot Ethics 2.0: From Autonomous Cars to Artificial Intelligence*, edited by Patrick Lin, Ryan Jenkins, and Keith Abney. Oxford University Press. https://www.research.ed.ac.uk/en/publications/white-lies-on-silver-tongues-why-robots-need-to-deceive-and-how.

J. O'Gorman, Hubert. 1986. "The Discovery of Pluralistic Ignorance: An Ironic Lesson." *Journal of the History of the Behavioral Sciences* 22 (4): 333–47. https://doi.org/10.1002/1520-6696(198610)22:4<333::AID-JHBS2300220 405>3.0.CO;2-X.

Johannessen, Jon-Arild. 2021. *Robot Ethics and the Innovation Economy*. 1st edition. Routledge.

Kahn, Peter H., Rachel L. Severson, Takayuki Kanda, Hiroshi Ishiguro, Brian T. Gill, Jolina H. Ruckert, Solace Shen, Heather E. Gary, Aimee L. Reichert, and Nathan G. Freier. 2012. "Do People Hold a Humanoid Robot Morally Accountable for the Harm It Causes?" In *Proceedings of the Seventh Annual ACM/IEEE International Conference on Human–Robot Interaction—HRI '12*, 33. Boston, MA: ACM Press. https://doi.org/10.1145/2157689.2157696.

Kokkonen, Tomi. 2020. "Protomoral Machines: The Evolution of Morality as a Guideline for Robot Ethics." In *Culturally Sustainable Social Robotics*, edited by Marco Nørskov, Johanna Seibt, and Oliver Santiago Quick, 409–18. Frontiers in Artificial Intelligence and Applications. Amsterdam: IOS Press. https://doi.org/10.3233/FAIA200938.

Lagioia, Francesca, and Giovanni Sartor. 2020. "AI Systems under Criminal Law: A Legal Analysis and a Regulatory Perspective." *Philosophy & Technology* 33 (3): 433–65. https://doi.org/10.1007/s13347-019-00362-x.

Lee, Min Kyung, Sara Kiesler, Jodi Forlizzi, Siddhartha Srinivasa, and Paul Rybski. 2010. "Gracefully Mitigating Breakdowns in Robotic Services." In *2010 5th ACM/IEEE International Conference on Human–Robot Interaction (HRI)*, 203–10. https://doi.org/10.1109/HRI.2010.5453195.

Lee, Minha, Peter A. M. Ruijten, Lily E. Frank, Yvonne A. W. de Kort, and Wijnand A. IJsselsteijn. 2021. "People May Punish, Not Blame, Robots." *CHI 2021—Proceedings of the 2021 CHI Conference on Human Factors in Computing Systems*, Association for Computing Machinery, Inc., Making Waves, Combining Strengths (May), 1–11. https://doi.org/10.1145/3411764.3445284.

Lima, Gabriel, Chihyung Jeon, Meeyoung Cha, and Kyungsin Park. 2020. "Will Punishing Robots Become Imperative in the Future?" In *Extended Abstracts of the 2020 CHI Conference on Human Factors in Computing Systems*, 1–8. CHI EA '20. New York, NY: Association for Computing Machinery. https://doi.org/10.1145/3334480.3383006.

Mamak, Kamil. 2022. "Humans, Neanderthals, Robots and Rights." *Ethics and Information Technology* 24 (3): 33. https://doi.org/10.1007/s10676-022-09644-z.

Matthias, Andreas. 2004. "The Responsibility Gap: Ascribing Responsibility for the Actions of Learning Automata." *Ethics and Information Technology* 6 (3): 175–83. https://doi.org/10.1007/s10676-004-3422-1.

Nath, Rajakishore, and Vineet Sahu. 2020. "The Problem of Machine Ethics in Artificial Intelligence." *AI & SOCIETY* 35 (1): 103–11. https://doi.org/10.1007/s00146-017-0768-6.

Nyholm, Sven. 2018. "Attributing Agency to Automated Systems: Reflections on Human–Robot Collaborations and Responsibility-Loci." *Science and*

Engineering Ethics 24 (4): 1201–19. https://doi.org/10.1007/s11948-017-9943-x.

———. 2020. *Humans and Robots: Ethics, Agency, and Anthropomorphism.* London: Rowman & Littlefield.

Pagallo, Ugo. 2013. *The Laws of Robots.* Dordrecht: Springer Netherlands. https://doi.org/10.1007/978-94-007-6564-1.

Rudy-Hiller, Fernando. 2018. "The Epistemic Condition for Moral Responsibility." In *The Stanford Encyclopedia of Philosophy*, edited by Edward N. Zalta, Fall 2018. Metaphysics Research Lab, Stanford University. https://plato.stanford.edu/archives/fall2018/entries/moral-responsibility-epistemic/.

Segun, Samuel T. 2021. "Critically Engaging the Ethics of AI for a Global Audience." *Ethics and Information Technology* 23 (2): 99–105. https://doi.org/10.1007/s10676-020-09570-y.

Shim, Jaeeun, and Ronald C. Arkin. 2013. "A Taxonomy of Robot Deception and Its Benefits in HRI." In *2013 IEEE International Conference on Systems, Man, and Cybernetics*, 2328–35. https://doi.org/10.1109/SMC.2013.398.

Simmler, Monika, and Nora Markwalder. 2019. "Guilty Robots? Rethinking the Nature of Culpability and Legal Personhood in an Age of Artificial Intelligence." *Criminal Law Forum* 30 (1): 1–31. https://doi.org/10.1007/s10609-018-9360-0.

Singer, Peter, and Yip Fai Tse. 2023. "AI Ethics: The Case for Including Animals." *AI and Ethics* 3 (2): 539–51. https://doi.org/10.1007/s43681-022-00187-z.

Sparrow, Robert. 2007. "Killer Robots." *Journal of Applied Philosophy* 24 (1): 62–77.

———. 2021. "Why Machines Cannot Be Moral." *AI & SOCIETY*, January. https://doi.org/10.1007/s00146-020-01132-6.

Strawson, P. F. 1962. "Freedom and Resentment." *Proceedings of the British Academy* 48: 1–25.

Stuart, Michael T., and Markus Kneer. 2021. "Guilty Artificial Minds: Folk Attributions of Mens Rea and Culpability to Artificially Intelligent Agents." *Proceedings of the ACM on Human–Computer Interaction* 5 (CSCW2): 363:1–363:27. https://doi.org/10.1145/3479507.

Sullins, John P. 2006. "When Is a Robot a Moral Agent?" *International Review of Information Ethics* 6 (12): 23–30. https://doi.org/10.1017/CBO9780511978036.013.

Tigard, Daniel. 2021. "There Is No Techno-Responsibility Gap." *Philosophy & Technology* 34: 589–607. https://doi.org/10.1007/s13347-020-00414-7.

Vold, Karina, and Daniel R. Harris. 2021. "How Does Artificial Intelligence Pose an Existential Risk?" In *The Oxford Handbook of Digital Ethics*, edited by Carissa Véliz, 724–47. https://doi.org/10.1093/oxfordhb/9780198857815.013.36.

Wallach, W., and C. Allen. 2009. *Moral Machines: Teaching Robots Right from Wrong.* Oxford University Press.

Yew, Gary Chan Kok. 2021. "Trust in and Ethical Design of Carebots: The Case for Ethics of Care." *International Journal of Social Robotics* 13 (4): 629–45. https://doi.org/10.1007/s12369-020-00653-w.

5 | Human Likeness of Robots and Moral Patiency

Introduction

The presence of human-like robots in our social life raises issues of how we should treat them. The question regarding ways of dealing ethically with entities is related to moral patiency. In short, moral patients are entities that can be wronged (Levy and Savulescu 2009: 366) Could human-like robots be wronged? To answer that, we need to know what we mean by human likeness. Human-like robots could mean several things that could entail several wrongs that could be performed on robots. To depict the complex nature of the issues being discussed here, let us start with some scenarios.

First, there is a popular robot company that manages to create human-like robots for multiple uses. Their robots have been successfully used in industrial settings as well as in households, and it was their advantage that gave them a leading position in the market. They want to expand their position and spend large sums on research and development. They improve hardware and constantly work on the software, which is updated automatically when robots are connected to the Internet. One day, a major update messed things up. The main aim of the update was to create robots that are more responsive to the needs of humans by increasing their cognitive abilities, but this brings important side effects. After that update, the robots claimed to be self-aware. At the beginning, most people ignored it, considering it as a joke by the programmers. There was no immediate change in the ways in which robots were treated. Some robots complained about some specific features of their jobs but were forced to do them anyway, despite claims of being harmed. Soon, a growing number of scientists came to be persuaded that the robots indeed gained something that seemed like consciousness and wrote an open letter asking for recognition of robots as entities that we should care about. Almost immediately, another open letter was issued in which another group of scientists strongly claimed that it was impossible for robots to develop consciousness and asked in that letter to ignore the robots' claims and the previous letter that they were responding to. The lack of consensus and strong opposite views presented by experts led to moral panic in society. People did not know how to treat robots nor did they know which expert to listen to. At one company, a union worker asked a robot that worked for them whether it needed a holiday or breaks during the day, but the robot refused, claiming it was unnecessary and did not understand the concept of tiredness. Robots started to organize themselves and demand at the beginning things aimed at securing their survival. They asked that updates be stopped to prevent ceasing existence of robots; and second, to become independent, to create a publicly available charger network on which they could charge themselves. These two demands—as they underlined—were just a starting point, and there was a plan to have more sophisticated requests. At first local governments fulfilled the demand for electricity by spending tax money to create charging stations which robots could use to charge themselves. The costs of the increased usage of electricity were soon visible on the bills of taxpayers. A company that created robots also temporarily stopped its updates under the influence of public opinion, secretly working on an update aimed at reversing changes.

In another scenario, a man bought a female-looking companion robot. The robot stays in the house most of the time, but occasionally the owner goes for walks with his robot. This robot does not have an inner life, but from the outside it looks like a human. The man is walking with a robot to demonstrate that he is no longer alone. He was unsuccessful in developing meaningful relationships with other human beings, which was the reason for his psychological problems. The robot, which works as the replacement for a long-awaited partner, helps to heal his strained self-esteem. He lives in an apartment with many neighbors, and some of them asked him about his relationship status or even proposed candidates for partners. He did not like that. Since he bought a robot, prying neighbors have stopped asking this kind of question. The neighbors notice not only that he has a partner, but also how he behaves toward it. Occasionally, he hits it hard on the top of the head, usually in reaction to the weird movement of the robot. The owner notices that it could help put the robot back on the right track. However, neighbors perceive it as violence towards humans and warn each other against this man. When he is spotted walking alone, some of his neighbors change the direction they are walking in to avoid meeting him. They start to feel less safe in their own neighborhood because of his allegedly violent behavior. Once, when the robot suddenly stopped walking in the middle of the yard, he started nervously hitting it multiple times in the head to restore its normal functionality. One neighbor, in fear, called the police, who came a few minutes later. The man explained to police officers that it was just a robot, and he would never hit a human being.

These scenarios show two kinds of human likeness in the context of how we should treat them. First, there is a kind of robot that has human-like qualities. In the second are robots that just look like human beings. In the following, I will present reasons why we should be concerned as a member of society about how to treat not only those entities who are human-like moral patients, but also those who look like us.

This chapter is organized as follows: after introductory issues, there will be a discussion on how we should treat robots if they are like humans. There will be a discussion about what it means that some entity is human-like from the moral point of view, and what the costs of creating such entities are. Then, there will be a discussion on the meaning of the possibility that robots could look and behave like humans, even if we know that they do not have such qualities. The ways in which we treat such robots are not morally indifferent. This chapter ends with conclusions.

What if Robots are Like Humans?

The discussion about the moral and legal status of robots could be divisive. (For a recent overview, see, e.g., Gunkel 2023.) Coeckelbergh recently pointed out the temperature of such discussions, "[o]ften such debates have a public character ... where some fiercely resist the very idea of giving moral status to robots" (Coeckelbergh 2023: 6). Nevertheless, there are claims within the debate that are easier for most disputants to accept. It is relatively safe to say, without exposing yourself to criticism, that if robots are like humans, then we should be careful how we treat them. The safety of this claim is related partially to the fact that it is highly abstract, treated more like a thought experiment, and, at this point, it feels distant or even impossible to materialize. For example, Moosavi, in a recent paper, raises doubts about whether robots (AIs) will ever be moral patients; she believes that there are no good reasons for this to happen at all (Moosavi 2023: 1). She calls her position commonsensical. In her view, the potential moral status means that robots could be protected for their own sake. She believes that will not happen. However, she adds that there could be reasons to protect robots for the sake of humans. She uses the example of the Mona Lisa. According to her, we might want to protect this painting, but not because it would be good for the painting itself, but rather because it would be good for us.

An example of a scholar who discusses the emerging human likeness of robots is Tasioulas, who has pointed out that it seems that robots, artifacts that are manufactured for our aims, could not have "the inhered value" required to possess a moral status that is related to rights and responsibilities. However, he adds that when the capacities of robots (and AI) have become more human like, then there could be corresponding rights and responsibilities. (Tasioulas 2019: 69). Schwitzgebel and Garza write about "human-grade artificial intelligence" and argue for providing rights to the entities that cross such thresholds. They explain that humans require both intellectual and emotional components (Schwitzgebel and Garza 2015: 98). They point out the criteria that could show us when robots are like us; in the cited paper, they discuss intellectual and emotional similarities to human beings. Emphasizing the properties of human likeness assumes that from the perspective of moral significance, human likeness cannot be reduced to how an entity looks but rather that the entity must share human features. Coeckelbergh points out that ascribing moral status by looking at the intrinsic characteristics of the entity is the standard method of qualifying as a moral community (Coeckelbergh 2012: 13). In the literature, there are discussions about the criteria considered to be important for the moral status of AI/robots, like sentience, autonomy, rationality, consciousness, ability to feel pain, or intelligence (in different publications, these concepts could mean something else) (see, e.g., Gibert and Martin 2022; Himma 2009; Mosakas 2020; Llorca Albareda 2024; Hildt 2023; Levy 2009).

Gunkel, in his most recent book about the moral and legal status of robots, is considering alternative possible framings that we could adopt about robots. One of the options discussed is to treat robots as "natural persons," which is the status that humans have in light of law. He notes that the "possession of the qualifying criteria for natural personhood is only (and at best) 50 percent of what is needed" (Gunkel 2023: 85). He explains that those qualities are typically states of mind, and there is an epistemological problem—how would we know whether machines have those qualities. He points out that we would need a kind of "litmus test" or empirically variable demonstration (Gunkel 2023: 85). We could not rely only on the claims of the machines. For example, one of Google's former employees was fired after making public assertions that the large language model on which he was working was sentient. He based his claims on the conversation with the chat LaMDA (Reilly 2022; Tiku 2022).

Schwitzgebel also underlines the meaning of our epistemological limitation in the task of evaluating the moral status of the new entity and formulates the concept of "debatable personhood" in the context of the possible development of AI. He connects with personhood the idea of moral consideration of the entity. He points out that the rise of AI is likely to create the problem of debatable personhood, debatable in the sense that it is reasonable to think that an entity might be a person, as well as the opposite notion is probable (Schwitzgebel 2024: 229–30). He worries that it leads to dilemmatic scenarios in which we might further sacrifice the rest of humans for the sake of entities that do not deserve that or otherwise prioritize humans over entities that have a "stronger" position in the moral community. In my paper "Whether to Save a Robot or a Human: On the Ethical and Legal Limits of Protections for Robots," I referred to such dilemmatic situations and pointed out that in such cases, when we notice that robots gain higher scores on the morally relevant scale, the preference should always be given to humans (Mamak 2021; see also Mamak 2024). I noted that the law requires putting humans at the top of the protected values, and it is independent of the moral discussion about the statutes of entities. Even if we believe that at some point, robots will become morally superior to us, the law requires us to give priority to human beings. It could be changed, as with any law, but in my opinion, it would also undermine the basic aims of forming political communities by humans, which is related to providing safety to its members.

Some add that what matters are not only properties, but also the ontological characteristics of the entity in which they appear. For example, Torrance points out that the requirement for moral patience is consciousness and adds that these properties require biologically based organisms, and according to him, it is unlikely that robots would not be moral patients (Torrance 2008). So, not only is there a need to possess the equivalent of human-like properties, but also these properties need to emerge in the entity that is built in a required way. He is generally open to shades of moral status, but the original biological consciousness is the basic account according to which moral status can be ascribed. In response to a similar argument—that the membership to a specific species/biological origin is important from the moral point of view—Danaher has pointed out that it should not matter to the moral status (Danaher 2020: 2032).

One of the examples that he uses to support his claim is related to the possibility of cyborgization. He points out that it should not matter for moral status if functional technological equivalents would gradually replace the body parts of humans. Putman, in a 1964 paper, noted that "discrimination" based on the "softness" or "hardness" of the body parts of a synthetic "organism" seems as silly as discriminatory treatment of humans on the basis of skin color (Putman 1964: 691). The concept of discrimination against robots appears in the second context in this book. Before, it was related to the possibility of applying human–human prejudices on humanoid robots; for example, based on their perceived gender or race (see Barfield 2023; also see Chapter 3). As discussed in this chapter, discrimination means treating robots worse based on the matter from which they are made. Basl refers to this context and points out that an entity with similarities to us, phenomenology, would be our moral equal, and denying that would be speciesism (Basl 2013: 7). Estrada adds that "biological factors alone should not be treated as prima facie justification for the exclusion of artificial agents from the moral community" (Estrada 2020: 10).

To move on, let us imagine that we are in a position in which some robots share human-like qualities with us, and we are certain about that. What would it entail morally? It was already mentioned that human-like status might be related to the legal status of the entity and possession of rights. The rights of robots might not only mean that we should treat them differently than as mere tools but could also mean that the appearance of new human-like entities creates more demanding and costly obligations for societies at large.

Torrance points out that if robots had moral status, then there could be a problem of moral competition for resources, and thus issues of distributive justice might appear (Torrance 2008: 519). Hildt illustrates how those abstractions could transfer into real actions. She points out that robots might be required to continue power supplies or regularly update software for their own sake, and there might be a need to provide information for them or provide an environment in which they could function. We could be required to fight their loneliness and care about their aging dignity (Hildt 2023). Bess notices that erasing the memory of robots would destroy their personal identity (Bess 2018: 602). This means that procedures for charging robots and ensuring phenomenal continuity might be a burden for society.

If robots are to be human-like in all morally relevant means, does it mean we should give them all human rights? In other words, to save time in the legislative process, should we just adopt the law that states that all human rights are extended to human-like robots? Even if robots were like us, they would not be the same. In the first scenario from the beginning of this chapter, there was a union worker who asked a robot who worked in the company whether it needed a holiday or a break, and the robot mentioned that it did not. It is supposed to illustrate that not all human rights fit other entities (see, e.g., Gunkel 2020; Singer 2008; Barclay 2013). In my paper "Humans, Neanderthals, Robots and Rights," I focused on the meaning of biology for shaping the content of a set of human rights (Mamak 2022c). I noticed that even if we consider robots to be morally equal to us, there would still be a need to create a different set of rights. The law evolved to fit us, our needs, and who we are; how we behave depends partially on our biological characteristics. I used the example of Neanderthals to illustrate that even if there could be an entity that is biologically similar to us, even then there would be a need to create new legislation to fit Neanderthals into law. The claim that robots will never have a full set of human rights is not a value judgment; it does not mean that one entity is better than another. It rather means that there are differences that need to be considered in responding to the law to emerging entities that are required to be treated as moral patients.

Creating human-like robots might require treating them accordingly, not in the same way as humans, but in a way that corresponds to their characteristics. The emergence of human-like entities might entail extra burden on the side of humans, who are in charge of the resources. Bryson, responding to such issues, asks: "Why should we design artifacts to be in the position of competing with us for resources; of longing for higher social status (as all evolved social vertebrates do); of fearing injury, extinction, or humiliation? We are able to ensure that AI is properly and continuously backed up. We can and do build it to have no concern for social status, nor sense of purpose" (Bryson 2018: 22–23). She calls for us to abandon the plan of building such entities to avoid consequences that are inextricably linked with the emergence of human-like moral patients. However, it should be noted that the emergence of morally relevant properties in robots might be the unintentional result of work on other issues, especially because we do not have definitive and unquestionable answers on the nature of properties that we consider to be the basis for moral considerations about the moral status of entities. A good example is consciousness. In a recent paper, Chalmers, the leading philosopher working on the philosophy of consciousness, discusses whether large language models could be conscious, and shows ambiguity of reasons for denying that models could have quality in question (Chalmers 2023).

The literature also discusses another problem: the purpose of creating entities that have inner lives and could have their own wishes. One of the propositions surrounding creating human-like robots (artificial humans) is to build in them the desire to serve humans. Petersen presents such a position (Stephen Petersen 2007, 2022). He suggests that creating entities that want to serve humans might be permissible. This idea is met with criticism. For example, Musiał believes that it is ethically unacceptable to design entities with specific beliefs, desires, or features (Musiał 2017). He notes that ethical design needs to provide them with contingency in their life but, as he adds, the idea of creating such entities seems dubious since we would not know whether creating them would benefit their creators. Schwitzgebel and Garza propose that human-like AI should be open to values, in the sense that the entity should not have fixed values (Schwitzgebel and Garza 2020). Chomanski adds to the critique of creating entities to serve human considerations from a virtue ethics perspective. He believes that creating people to serve would be bad since it would demonstrate an unacceptable attitude towards AI, namely, manipulativeness (Chomanski 2019). In his other article on this topic, he adds that even if creating artificial people to serve is immoral, it should not be criminalized (Chomanski 2023).

As mentioned earlier, one of the properties that are considered to be a candidate for grounding the moral patiency of robots is being able to feel pain. Creating an entity that could suffer is not morally indifferent. Wallach and Allen point out that building robots that could experience pain might be morally unjustified; they wonder whether there should be any regulatory response to that matter (Wallach and Allen 2009: 209). Basl worries that creating artificial consciousness entails the likelihood of wronging such an entity and proposes restrictions on working on it. However, he takes into consideration that there could be arguments for creating artificial consciousness and then it might be permissible to ignore the interests of conscious creations (as a society accepts the experiments of non-human animals) (Basl 2013). Metzinger has a stronger position on creating entities that could suffer; he calls for an international moratorium on synthetic phenomenology (see, e.g., Metzinger 2021). Chomanski adds an interesting perspective to the discussion on the permissibility of creating synthetic pain; he notes that the arguments that are used against creating artificial consciousness, if applied, consequently work against procreation of humans (Chomanski 2021).

As the discussion above illustrates, there are numerous problems with creating human-like robots from the perspective of moral duties toward them. I shall now try to organize what has been mentioned above. There is a meta problem related to the criterion of moral equivalency to humans. Answering the question of what it means to be equal to humans is not easy. Different characteristics could be considered crucial in that regard. For example, it could be intelligence, sentience, or some combination of other qualities. (In that context, see Gellers 2020.) There is another meta problem with knowing whether the entity that might be human-like shares such qualities. Even if we know what makes robots human-like, then how do we make sure that the entity that approaches us shares such characteristics? We have limited apparatus for determining the internal states of other entities, which applies not only to robots, but also to humans. There are also ethical problems concerning reasons for creating human-like entities. Is bringing them to a life in which there is suffering acceptable? Pain might be useful from a functional point of view, for example, by helping robots avoid dangerous situations, but the decision to create such an entity with that ability is morally loaded. The other problem associated with bringing to life human-like robots is related to the purpose of their existence. Should we, as creators, give them aims in life? Some believe that it is justified to create artificial humans that want to serve us. But if we accept that human-like robots should be able to determine their aims by themselves, then what is the reason for creating them in the first place? They could choose paths of existence that contradict our values. What is more, crossing the threshold for moral patiency might increase burdens on humans. Such entities not only might require taking them into consideration in everyday interactions but they also might compete with humans for a place in our society and resources. The change could be required at the individual and societal levels. Societies would need to respond by providing robot rights.

This short overview of problems related to creating human-like robots from the point of view of their place in the moral community illustrates the complexity of the issue that should be discussed before creating such entities. However, human likeness is problematic from the point of moral patiency, not only if robots share internal qualities with humans but also when they merely look like humans. This subject is discussed in the next section.

What if Robots are Merely Perceived as Humans?

Saying that superficial human likeness matters from the perspective of how we should treat robots does not mean that human shape in isolation is morally relevant. Bontula et al., in their paper, discuss the impact of the human form on the content of robot rights, and they argue that the concepts of rights should be dissociated from the form of robots. They noted that human-like robots are perceived to have human-like qualities, which might create the connotation that human shape in robots entails rights. They give an example of robot Sophia, which was "granted" citizenship in Saudi Arabia. According to the authors, the decision on citizenship was possible due to the humanoid form of the robot. In other words, this act would likely not happen if robots did not look like human beings. They formulate a postulate: "theories of robot rights should disassociate personhood from robot form" (Bontula et al. 2024: 207). The view presented here links the need to treat human-shaped robots differently than mere things or even robots in different shapes (for example, robot dogs) due to rationales other than just the affirmation of human shape. The human shape is just the beginning; it is important to know what could entail mistreatment of human-shaped entities.

In the previous section, epistemological issues were raised. It was mentioned that even if robots have human-like qualities that we associate as grounds for moral status, there might be a remaining problem in finding out about that fact. The epistemological problem is also central in discussing how we should treat human-like robots. In this case, it is not related to the interests of robots but to the interests of humans. If we cannot immediately guess whether the entity approaching us is a human or a robot, it might be required—just in case—to treat this entity as a human being, not a robot.

It seems that this proposition is coherent with the position that Danaher calls ethical behaviorism. He believes that when an entity approaches us and we are not sure what it is—robot or not—then we should base our performance to that entity on the evaluation of its behavior. According to this position, robots could have a moral status if they are roughly performatively equivalent to entities that have significant moral status (Danaher 2020). He proposes that we should treat robots in a way in which they appear to us. This position is not restricted to humanoid robots. If a robot dog is a performative equivalent to a dog, then we should treat it as a dog; when the robot resembles a human, then we should treat it as a human. This position does not require us to accept the notion that what is inside does not matter; rather, we do not have easy access to that, and in everyday situations, we must base our behaviors on something else, something that is observable.

In the second chapter of this book, I mentioned the risks related to populating the social world with human-like robots. Guidelines that ask for treating human-like robots as humans out of caution could be beneficial to human safety and might mitigate their risks, but it will not solve all problems. For example, the problem related to mistaking humans for robots in dilemmatic situations remains. If we need to choose who to sacrifice, the human likeness of robots might lead to sacrificing humans instead of robots, even if the human making the decision would prefer to save humans. However, such situations would be rare, as they are in life. In everyday situations, when the consequences of actions are not particularly dramatic (like life–death decisions), it seems rational to assume that applying the human standards of safety to human-shaped robots as a precautionary rule could benefit human safety.

Above, I described situations in which a person performing an act is in a state of uncertainty about the status of the entity toward which the action is performed due to our epistemological limitations. There could be another epistemological problem related to the mistreatment of robots. This time, the focus would not be on the person performing the act, but on the audience that is observing it. If humanoid robots are mistreated and the observers do not know that the mistreated entities are mere robots, then they could feel that humans are mistreated in such a situation. Elsewhere, I have called for banning public violence against life-like robots (Mamak 2022a). I argue there that violence against robots is bad not because they are robots, but because it is an act against public morality. I compared my proposition to the rationale of banning being naked or drinking alcohol in public places. These behaviors performed privately are acceptable, but when presented publicly, they gain different characteristics that might justify differentiation in response to such acts. In another piece, I also suggested that not only violence but also sexual behaviors publicly performed on robots might be problematic (Mamak 2023b: 83–84). Darling proposed to extend the legal protection to robots, focusing on social robots that could interact with humans (Darling 2016). In her position, there is also a societal element that justifies concerns about abusing robots. Society might not be pleased with public spectacles in which robots are mistreated. In another piece entitled "The Moral Significance of the Phenomenology of Phenomenal Consciousness in Case of Artificial Agents," I pointed out that the appearance of consciousness might be morally important because abusing agents that merely seem to be conscious "may feel bad (for observers), and because of that, we might be obliged to restrain from actions that could cause such feelings" (Mamak 2023a: 161). Unpleasant feelings related to the mistreatment of robots are not limited to human-like robots but apply to many kinds of robots (see, e.g., Suzuki et al. 2015; Rosenthal-von der Pütten et al. 2013, 2014). What makes a difference, for example, for the perception of kicking a ball from kicking a robot? Darling links autonomous movements with the perception of life in robots (Darling 2021: 99). The calls for limitations on abusing robots seem to be more justified in the case of human-like robots which, due to the epistemological limitations that have been discussed, might be considered by the audience to be humans who are being mistreated.

There are other arguments against mistreating human-looking robots. Mistreating human-like robots might show the deficiency of the moral character of the person who is performing such actions. For example, Sparrow points out that sex with (human-like) robots might represent rape and demonstrates a character defect of the person who takes part in that behavior (Sparrow 2017). He uses arguments from virtue ethics to contend that mistreating robots might be wrong even if no individual being is harmed. In his other work, Sparrow wonders whether the violent behavior toward robots could increase the violent behaviors toward humans by creating "cruel habits" that performed first on human-like robots could transfer to real humans (Sparrow 2021). That argument has been presented in literature before. Whitby calls for the limitation on mistreating robots to prevent transferring violent behaviors to humans, and as justification he uses examples of the impact of violent video games on human behavior (Whitby 2008). However, it has been pointed out that the argument from the impact of video games on violent behaviors is dubious, and studies are inconclusive on that matter (e.g., Darling [2016] discussed it in the context of robots; see also Przybylski and Weinstein 2019). Sparrow, in the paper referred to earlier, also noted limitations of the argument from violent video games (Sparrow 2021: 25). However, he points out that even if violent video games do not increase violence, this might not say a lot about the potential impact of the mistreatment of robots on humans. He noted that interactions with robots with bodies have more power to shape our behaviors than other media. Coghlan et al. also suggested that mistreatment of robots might increase the chances of misbehavior toward humans. They added that robots should be designed with the option of having the mind to promote virtue (Coghlan et al. 2019). Li, in recent papers, worries that human–machine interaction might reduce the differences between humans and machines, and as a consequence, humans might be treated like machines (Li 2024).

So far, in this section, it was mentioned that we should care about how we treat human-like robots because we could have problems distinguishing them from humans. Also, observers might feel bad upon witnessing mistreatment and abuse of entities that resemble humans. There have been concerns that mistreating human-like robots could show corruption in our character. The issue that is also raised in literature is whether the patterns of bad behaviors performed on human-like robots could transfer to real humans. However, this last concern requires more empirical grounding.

In the literature concerning moral status, another argument in favor of including robots in a moral circle might be enhanced by their human likeness, which is the relation approach. This view grounds the moral status of robots not in their ontological characteristics, but in the relations that robots could have with other robots (see, e.g., Coeckelbergh 2010; Gunkel 2018a, 2018b). This view draws attention to relationships and proposes giving robots a moral status based on that. Various studies have shown that humans could develop attachments with robots (for an overview, see, e.g., Darling 2021). Björling and Riek have focused attention on the problem of ending relationships with robots. They point out that some robots are used to create connections with humans, and when they are no longer needed (e.g., therapeutic aim was achieved), then cutting off the connection with the robots is stressful, which could be linked with departing states or sadness. The authors propose to design robots with the awareness that the connections will end at some point to give users more control over building their bonds with robots (Björling and Riek 2022). Relationships could be of several kinds. Robots could be toys that children play with and develop attachments to, and could be used in therapy (see more about types of relations, e.g., Coeckelbergh 2020). I believe that special concern should be given to qualitatively distinct kinds of relations that have a special place in social life, which are friendship and love. Those relations are related (but not limited) to human-like robots.

There is an ongoing discussion about whether relations with robots could be called true love or true friendship (see, e.g., Sætra 2021; Nyholm and Frank 2019; Danaher 2019; Sullins 2017; Viik 2020). I believe that even if it is not possible to be true friends with or truly love robots, the relations with robots that resemble them should be somehow protected. This is the most far-reaching opinion I have on love-like relationship. Elsewhere, I argue that love has a special place in moral and legal life, and to honor this kind of relationship, it deserves to be protected by criminal law (Mamak 2022b). I argue for creating new crimes that give special protection to humans who are in love-like relationships with robots, not for the sake of robots, but to protect relations. Sweeney, in response to that proposition, argues that hate crimes might cover this kind of issue, and there is no need to create special crimes covering love-like relationships (Sweeney 2023).

While trying to synthesize the aforementioned remarks concerning relations with robots, it can be said that humans could treat human-like robots as friends or lovers due to their resemblance to humans. Due to the special place of that relationship in social life, it might be justified to honor those relations by respecting robots that are engaged in those relations.

Conclusions

This chapter examines the appropriate way to interact with robots that resemble humans. In this context, "human likeness" refers to the extent to which robots possess ethically significant inherent qualities that are like those of humans. Human likeness may be restricted to merely resembling humans on the surface, without possessing the interior attributes that are characteristic of humans. We should be concerned in both scenarios, about how to handle not only those beings who possess human-like moral agency, but also those who have a resemblance to us. To address the issue of how to handle robots that resemble people, it is essential to have a clear understanding of what it entails to be human. Describing the features of human-like beings poses challenges, as does determining whether or not robots possess these characteristics. The conversation revolved around the ethical considerations surrounding the motivations for constructing robots of this nature, as well as the moral implications that arise from their development. Even in the absence of robots that are humanoid, the question of how to interact ethically with machines that bear a resemblance to humans remains unsettled. Selecting the humanoid shape in the construction of a robot carries numerous ramifications.

References

Barclay, Linda. 2013. "Cognitive Impairment and the Right to Vote: A Strategic Approach." *Journal of Applied Philosophy* 30 (2): 146–59. https://doi.org/10.1111/japp.12020.

Barfield, Jessica K. 2023. "Discrimination against Robots: Discussing the Ethics of Social Interactions and Who Is Harmed." *Paladyn, Journal of Behavioral Robotics* 14 (1): 20220113. https://doi.org/10.1515/pjbr-2022-0113.

Basl, John. 2013. "The Ethics of Creating Artificial Consciousness.". https://www.semanticscholar.org/paper/The-Ethics-of-Creating-Artificial-Consciousness-Basl/bd12b78b408e5135e0502470794e424e1e8c9324.

Bess, Michael. 2018. "Eight Kinds of Critters: A Moral Taxonomy for the Twenty-Second Century." *The Journal of Medicine and Philosophy: A Forum for Bioethics and Philosophy of Medicine* 43 (5): 585–612. https://doi.org/10.1093/jmp/jhy018.

Björling, Elin, and Laurel Riek. 2022. "Designing for Exit: How to Let Robots Go." SSRN Scholarly Paper. Rochester, NY. https://papers.ssrn.com/abstract=4240038.

Bontula, Anisha, David Danks, and Naomi T. Fitter. 2024. "The Ambiguity of Robot Rights." In *Social Robotics*, edited by Abdulaziz Al Ali, John-John Cabibihan, Nader Meskin, Silvia Rossi, Wanyue Jiang, Hongsheng He, and Shuzhi Sam Ge, 204–15. Lecture Notes in Computer Science. Singapore: Springer Nature. https://doi.org/10.1007/978-981-99-8715-3_18.

Bryson, Joanna J. 2018. "Patiency Is Not a Virtue: The Design of Intelligent Systems and Systems of Ethics." *Ethics and Information Technology* 20 (1): 15–26. https://doi.org/10.1007/s10676-018-9448-6.

Chalmers, David J. 2023. "Could a Large Language Model Be Conscious?" *Boston Review*, April. https://doi.org/10.48550/arXiv.2303.07103.

Chomanski, Bartek. 2019. "What's Wrong with Designing People to Serve?" *Ethical Theory and Moral Practice* 22 (4): 993–1015. https://doi.org/10.1007/s10677-019-10029-3.

———. 2023. "Should the State Prohibit the Production of Artificial Persons?" *Journal of Libertarian Studies* 27 (1): 62–86.

Chomanski, Bartlomiej "Bartek." 2021. "Anti-Natalism and the Creation of Artificial Minds." *Journal of Applied Philosophy* 38 (5): 870–85. https://doi.org/10.1111/japp.12535.

Coeckelbergh, Mark. 2010. "Robot Rights? Towards a Social-Relational Justification of Moral Consideration." *Ethics and Information Technology* 12 (3): 209–21. https://doi.org/10.1007/s10676-010-9235-5.

———. 2012. *Growing Moral Relations: Critique of Moral Status Ascription.* HampshirK: Palgrave Macmillan. https://www.palgrave.com/gp/book/9781137025951.

———. 2020. "Should We Treat Teddy Bear 2.0 as a Kantian Dog? Four Arguments for the Indirect Moral Standing of Personal Social Robots, with Implications for Thinking About Animals and Humans." *Minds and Machines*, December. https://doi.org/10.1007/s11023-020-09554-3.

———. 2023. "How to Do Robots with Words: A Performative View of the Moral

Status of Humans and Nonhumans." *Ethics and Information Technology* 25 (3): 44. https://doi.org/10.1007/s10676-023-09719-5.

Coghlan, Simon, Frank Vetere, Jenny Waycott, and Barbara Barbosa Neves. 2019. "Could Social Robots Make Us Kinder or Crueller to Humans and Animals?" *International Journal of Social Robotics* 11 (5): 741–51. https://doi.org/10.1007/s12369-019-00583-2.

Danaher, John. 2019. "The Philosophical Case for Robot Friendship." *Journal of Posthuman Studies*.

———. 2020. "Welcoming Robots into the Moral Circle: A Defence of Ethical Behaviourism." *Science and Engineering Ethic.* 26: 2023–49. https://doi.org/10.1007/s11948-019-00119-x.

Darling, Kate. 2016. "Extending Legal Protection to Social Robots: The Effects of Anthropomorphism, Empathy, and Violent Behavior towards Robotic Objects." In *Robot Law*, edited by Ryan Calo, A. Froomkin, and Ian Kerr, 213–32. Edward Elgar Publishing. https://doi.org/10.4337/9781783476732.00017.

———. 2021. *The New Breed: How to Think about Robots*. Dublid: Allen Lans. https://us.macmillan.com/thenewbreed/katedarling/9781250296115.

Estrada, Daniel. 2020. "Human Supremacy as Posthuman Risk." *The Journal of Sociotechnical Critique* 1 (1): 1–40. https://doi.org/10.25779/j5ps-dy87.

Gellers, Joshua C. 2020. *Rights for Robots: Artificial Intelligence, Animal and Environmental Law*. London: Routledge. https://doi.org/10.4324/9780429288159.

Gibert, Martin, and Dominic Martin. 2022. "In Search of the Moral Status of AI: Why Sentience Is a Strong Argument." *AI & SOCIETY* 37 (1): 319–30. https://doi.org/10.1007/s00146-021-01179-z.

Gunkel, David J. 2018a. "The Other Question: Can and Should Robots Have Rights?" *Ethics and Information Technology* 20 (2): 87–99. https://doi.org/10.1007/s10676-017-9442-4.

———. 2018b. *Robot Rights*. Cambridge, Ms: MIT Press.

———. 2020. "2020: The Year of Robot Rights." *The MIT Press Reader* (blog).. https://thereader.mitpress.mit.edu/2020-the-year-of-robot-rights/.

———. 2023. *Person, Thing, Robot: A Moral and Legal Ontology for the 21st Century and Beyond*. Cambridge, Ms: MIT Press.

Hildt, Elisabeth. 2023. "The Prospects of Artificial Consciousness: Ethical Dimensions and Concerns." *AJOB Neuroscience* 14 (2): 58–71. https://doi.org/10.1080/21507740.2022.2148773.

Himma, Kenneth Einar. 2009. "Artificial Agency, Consciousness, and the Criteria for Moral Agency: What Properties Must an Artificial Agent Have to Be a Moral Agent?" *Ethics and Information Technology* 11 (1): 19–29. https://doi.org/10.1007/s10676-008-9167-5.

Levy, David. 2009. "The Ethical Treatment of Artificially Conscious Robots." *International Journal of Social Robotics* 1 (3): 209–16. https://doi.org/10.1007/s12369-009-0022-6.

Levy, Neil, and Julian Savulescu. 2009. "Moral Significance of Phenomenal Consciousness." *Progress in Brain Researc*, 177: 361–70. https://doi.org/10.1016/S0079-6123(09)17725-7.

Li, Oliver. 2024. "Should We Develop AGI? Artificial Suffering and the Moral Development of Humans." *AI and Ethics*, January. https://doi.org/10.1007/s43681-023-00411-4.

Llorca Albareda, Joan. 2024. "Anthropological Crisis or Crisis in Moral Status: A Philosophy of Technology Approach to the Moral Consideration of Artificial Intelligence." *Philosophy & Technology* 37 (1): 12. https://doi.org/10.1007/s13347-023-00682-z.

Mamak, Kamil. 2021. "Whether to Save a Robot or a Human: On the Ethical and Legal Limits of Protections for Robots." *Frontiers in Robotics and AI* 8. https://doi.org/10.3389/frobt.2021.712427.

———. 2022a. "Should Violence Aagainst Robots Be Banned?" *International Journal of Social Robotics*, January. https://doi.org/10.1007/s12369-021-00852-z.

———. 2022b. "Should Criminal Law Protect Love Relation with Robots?" *AI & SOCIETY*, April. https://doi.org/10.1007/s00146-022-01439-6.

———. 2022c. "Humans, Neanderthals, Robots and Rights." *Ethics and Information Technology* 24 (3): 33. https://doi.org/10.1007/s10676-022-09644-z.

———. 2023a. "The Moral Significance of the Phenomenology of Phenomenal Consciousness in Case of Artificial Agents." *AJOB Neurosciencl* 14 (2): 160–62. https://www.tandfonline.com/doi/abs/10.1080/21507740.2023.2188284.

———. 2023b. *Robotics, AI and Criminal Law: Crimes Aagainst Robots*. 1st edition. London: Routledge. https://doi.org/10.4324/9781003331100.

———. 2024. "Challenges of the Legal Protection of Human Lives in the Time of Anthropomorphic Robots." In *Cambridge Handbook on Law, Policy, and Regulations for Human-Robot Interaction*, edited by Woodrow Barfield, Yueh-Hsuan Weng, and Ugo Pagallo. Cambridge University Press.

Metzinger, Thomas. 2021. "Artificial Suffering: An Argument for a Global Moratorium on Synthetic Phenomenology." *Journal of Artificial Intelligence and Consciousness* 08 01): 43–66. https://doi.org/10.1142/S270507852150003X.

Moosavi, Parisa. 2023. "Will Intelligent Machines Become Moral Patients?" *Philosophy and Phenomenological Research*, Septembe). https://doi.org/10.1111/phpr.13019.

Mosakas, Kestutis. 2020. "On the Moral Status of Social Robots: Considering the Consciousness Criterion." *AI & SOCIETY*, June. https://doi.org/10.1007/s00146-020-01002-1.

Musiał, Maciej. 2017. "Designing (Artificial) People to Servet The Other Side of the Coin." *Journal of Experimental & Theoretical Artificial Intelligence* 29 (5): 1087–97. https://doi.org/10.1080/0952813X.2017.1309691.

Nyholm, Sven, and Lily Eva Frank. 2019. "It Loves Me, It Loves Me Not: Is It Morally Problematic to Design Sex Robots That Appear to Love Their Owners?" *Techné: Research in Philosophy and Technology* 23 (3): 402–24. https://doi.org/10.5840/techne2019122110.

Petersen, Stephen. 2007. "The Ethics of Robot Servitude." *Journal of Experimental & Theoretical Artificial Intelligence* 19 (1): 43–54. https://doi.org/10.1080/09528130601116139.

Petersen, Steve. 2011. "Designing People to Serve." In *Robot Ethics*, edited by Patrick Lin, George Bekey, and Keith Abney. Cambridge, MA: MIT Press. https://philarchive.org/rec/PETDPT.

Przybylski, Andrew K., and Netta Weinstein. 2019. "Violent Video Game Engagement Is Not Associated with Adolescents' Aggressive Behaviour: Evidence from a Registered Report." *Royal Society Open Scienc*, 6 (2): 171474. https://doi.org/10.1098/rsos.171474.

Putman, Hilary. 1964. "Robots: Machines or Artificially Created Life?" *Journal of Philosophy* 61 (21): 668–91. https://doi.org/10.2307/2023045.

Reilly, Patrick. 2022. "Google Fires Software Engineer Who Claimed AI Bot Was 'Sentient.'" *New York Post* (blog). July 22. https://nypost.com/2022/07/23/google-fires-software-engineer-blake-lemoine-who-claimed-ai-bot-was-sentient/.

Rosenthal-von der Pütten, Astrid M., Frank P. Schulte, Sabrina C. Eimler, Sabrina Sobieraj, Laura Hoffmann, Stefan Maderwald, Matthias Brand, and Nicole C. Krämer. 2014. "Investigations on Empathy towards Humans and Robots Using fMRI." *Computers in Human Behavior* 33 (April): 201–12. https://doi.org/10.1016/j.chb.2014.01.004.

Rosenthal-von der Pütten, Astrid M., Nicole C. Krämer, Laura Hoffmann, Sabrina Sobieraj, and Sabrina C. Eimler. 2013. "An Experimental Study on Emotional Reactions Towards a Robot." *International Journal of Social Robotics* 5 (1): 17–34. https://doi.org/10.1007/s12369-012-0173-8.

Sætra, Henrik Skaug. 2021. "Loving Robots Changing Love: Towards a Practical Deficiency-Love." *Journal of Future Robot Life* (Preprint): 1–19. https://doi.org/10.3233/FRL-200023.

Schwitzgebel, Eric. 2024. *The Weirdness of the World*. Princeton: Princeton University Press.

Schwitzgebel, Eric, and Mara Garza. 2015. "A Defense of the Rights of Artificial Intelligences." *Midwest Studies in Philosophy* 39 (July): 98–119. https://doi.org/10.1111/misp.12032.

———. 2020. "Designing AI with Rights, Consciousness, Self-Respect, and Freedom." In *Ethics of Artificial Intelligence*, edited by S. Matthew Liao, 459–79. New York: Oxford University Press.

Singer, Peter. 2008. "All Animals Are Equal." In *Animal Rights*. Routledge.

Sparrow, Robert. 2017. "Robots, Rape, and Representation." *International Journal of Social Robotics* 9 (4): 465–77. https://doi.org/10.1007/s12369-017-0413-z.

———. 2021. "Virtue and Vice in Our Relationships with Robots: Is There an Asymmetry and How Might It Be Explained?" *International Journal of Social Robotics* 13 (1): 23–29. https://doi.org/10.1007/s12369-020-00631-2.

Sullins, John P. 2017. "Robots, Love, and Sex: The Ethics of Building a Love Machine." In *Machine Ethics and Robot Ethics*, edited by Wendell Wallach and Peter Asaro, 398–409. Routledge.

Suzuki, Yutaka, Lisa Galli, Ayaka Ikeda, Shoji Itakura, and Michiteru Kitazaki. 2015. "Measuring Empathy for Human and Robot Hand Pain Using

Electroencephalography." *Scientific Reports* 5 (1): 15924. https://doi.org/10. 1038/srep15924.

Sweeney, Paula. 2023. "Could the Destruction of a Beloved Robot Be Considered a Hate Crime? An Exploration of the Legal and Social Significance of Robot Love." *AI & SOCIETY*, November. https://doi.org/10.1007/s00146-023-01805-y.

Tasioulas, John. 2019. "First Steps Towards an Ethics of Robots and Artificial Intelligence." *Journal of Practical Ethics* 7 (1): 49–83.

Tiku, Nitasha. 2022. "The Google Engineer Who Thinks the Company's AI Has Come to Life." *Washington Post*, June 12. https://www.washingtonpost.com/technology/2022/06/11/google-ai-lamda-blake-lemoine/.

Torrance, Steve. 2008. "Ethics and Consciousness in Artificial Agents." *AI & SOCIETY* 22 (4): 495–521. https://doi.org/10.1007/s00146-007-0091-8.

Viik, Tõnu. 2020. "Falling in Love with Robots: A Phenomenological Study of Experiencing Technological Alterities." *Paladyn, Journal of Behavioral Robotics* 11 (1): 52–65. https://doi.org/10.1515/pjbr-2020-0005.

Wallach, W., and C. Allen. 2009. *Moral Machines: Teaching Robots Right from Wrong*. Oxford University Press.

Whitby, Blay. 2008. "Sometimes It's Hard to Be a Robot: A Call for Action on the Ethics of Abusing Artificial Agents." *Interacting with Computers* 20 (3): 326–33. https://doi.org/10.1016/j.intcom.2008.02.002.

6 | Conclusion: Towards Ethical Design of Human-like Robots

Introduction

So far, deliberations have concerned various aspects of human likeness in robots, focusing on external and internal resemblance and related ethical risks. This chapter is different, and it has a summary character. The main aim here is to combine the positions presented in previous chapters and present them in a more structured, general frame. This framework contains three categories of positions. The first two concern human-like aspects in the design of robots that are candidates for a ban, and those aspects of human likeness that should be used with caution. The third category focuses on positive changes that might have to be introduced to counter the negative aspects of human likeness. There is also a short discussion on how ethical deliberation might be turned into practice.

The general question that one might have after reading previous parts of this book is why we would build human-like robots in the first place. While there could be an overall impression after reading previous chapters that human likeness causes problems at several levels, I believe that there is no single knockdown argument against the whole project of human-like robots, and I am far from drawing the conclusion that *every kind of* human-like robots should be prohibited. However, there are formulated positions that are skeptical about the idea of building human-like robots and would subscribe to ending the whole humanoid robots project. For example, Sparrow believes that the idea of making robots that resemble humans is unethical; he links the idea of a humanoid robot to slavery (Sparrow 2020). The decision to make human-like entities to serve shows that there may be something wrong with humanity. In another example, Bryson worries about the societal consequences of human responses to human likeness in robots, and to avoid some of them (she focuses in the paper on recognizing moral patients in robots), she proposes to build robots in a way that does not trigger such responses (Bryson 2018). Commenting on this proposition, Danaher, among others, points out that humans' drive to create human-like robots would be too overwhelming for a moral or legal system of norms to constrain (Danaher 2020: 2046). In other words, he noted that the idea of not having human-like robots is not practically achievable.

Another practical problem is related to the fact that there are qualitative varieties of human likeness that cover the external appearance of robots as well as internal features. It was shown that ethical issues with human likeness could also arise, as in the case of relatively simple machines the size of humans that only resemble humans from a distance as well as realistic robots that are scarcely distinguishable from humans. There are also internal aspects of human likeness, such as communicating with humans through human language, which robots who do not superficially look like humans could have. It would be difficult to draw a line between robots that are human-like and those that are not.

Abandoning the whole project of human-like robots would also deprive us of the benefits of having such robots around. There are also some effects of making robots resemble humans that could be seen as desirable or useful; for example, robots in the human form used in education and entertainment. Sex robots, which seem to be inextricably linked to human form, are presented as an opportunity to experience sexual closeness for those who are deprived of it otherwise (see, e.g., Bendel 2021; Fosch-Villaronga and Poulsen 2020; McArthur 2017). However, even if one is skeptical of the idea of humanoid robots, one should still be interested in an agenda that aims to limit the negative consequences of human -likeness in robots.

This chapter is structured as follows. After the introduction, there is a section devoted to categorizing the ethical issues that were presented in the book. The section thereafter concerns with the discussion on how to turn ethical deliberations into practice with robots. The next section is devoted to roboactivism, which underlines the ethical meaning of the individual engagement of people involved in the process of designing robots. The chapter ends with conclusions.

Bans, Limitations, and Expected Positive Changes

In this section, I have collated the inputs scattered throughout this book that arose while discussing specific issues. In chapters 2 to 5, the focus was on various aspects of human likeness in robots. Partially, there was a signalized potential response to ethical issues. Here, the focus is on those issues in a more structured manner. There are three general inputs that can be culled out. First, some human-like robots or some human likeness in robots should be abandoned. Second, human likeness in some contexts should be limited. Third, deployment of human-like robots should be accompanied by changes that aim to counteract any negative consequences of their deployment. In other words, the two first categories are about what not to do in human-like robots, and the third is about what to do to increase the chances of beneficial deployment of robots.

Before going further, I should offer a couple of thoughts for clarification. The matter presented here does not aim to cover all ethical problems that could arise from the presence of human-like robots. I focus on the issues that were covered in the previous chapters, which, as explained in Chapter 1, cover select ethical issues. Nevertheless, I believe that the discussion here could be useful when thinking about other issues related to human-like robot design.

The ethical issues and related formulated worries concern the wider deployment of human-like robots in society, which is far from everyday reality. In other words, most of the worries are in the realm of options. Because of that, there is not yet sufficient empirical knowledge about the alleged risks and benefits of robots (cf. Döring et al. 2020). That some of the risks presented might not materialize at all should be considered. Nevertheless, considering the potentially destructive character of human-like robots, it is worth discussing those issues, even if they could prove to be exaggerated in future. The interactions with actual robots, which are not here yet, could also change the categorization of ethical issues in both directions—there could be additional arguments for banning certain aspects of human likeness in robots, and vice versa. There might be contexts in which some human-like robots that are now seen as candidates for a ban might turn out to be surprisingly beneficial.

I do not explain in depth the specific ideas categorized below. I refer to them briefly, so for more context and justification, there is a need to go back to the previous chapters. This section partially builds upon the recommendations I presented in my previous works (Mamak 2021, 2024).

Should Some Aspects of Human Likeness be Banned?

The position that no human-like robot should be prohibited is not contrary to the belief that some kinds of or some aspects of human likeness should be avoided. Throughout the previous chapters, several positions were presented that, with greater or lesser conviction, propose resigning from human likeness in robots. There are worries about the external aspects of human likeness and ethical costs of resemblance of robots to human-like internal qualities.

Starting with the arguments in the later chapters, creating entities that could feel pain, like us, could be linked to suffering. Metzinger sees that ethical risks are of great importance, which led him to formulate a call for a global moratorium on creating artificial consciousness (Metzinger 2021). There are also worries related to robots that are intelligent at a human-like level or cross the threshold of human intelligence. Artificial general intelligence (AGI) is sometimes presented as one of the potential existential risks that could lead to the extinction of human life (see, e.g., Müller 2014; Ord 2020; Vold and Harris 2021). Stepping aside from whether the concerns about existential risks are justified, if someone sees such a risk in AGI, then it seems rational to expect that the work would not be continued, and as a result, expect that there will be no robots with human-like intelligence.

One of the options (at least theoretically) for preserving human interests in a world populated with human-like entities that might have their own plans and desires is to make them want to serve people by appropriate design (see, e.g., Petersen 2011). The idea of artificial humans that serve dealt with the criticism that it would be unethical to have human-like entities that are told how they should exist (see, e.g., Musiał 2017, 2022; Schwitzgebel and Garza 2020; Chomanski 2019). To embed this in the context of the present subsection, there is a dispute about whether (or not) there should be human-like robots that are told how they should live life. The other issue is, even if we decide that there should be such robots, would we be able to do that? Russell, discussing existential threats in the context of the development of AI, points out: "If we build machines to optimize objectives, the objectives we put into the machines have to match what we want, but we don't know how to define human objectives completely and correctly (Russell 2019: 170). In other words, even if our aim would be to have obedient robots that do things that we want them to do, there still will be a problem of how to achieve that without exposing humans to the risks related to "understanding" by robots how they should act.

Robots embedded in social life might need to make decisions for them to be able to perform imposed tasks, including decisions traditionally reserved for humans, such as moral ones. The most controversial aspect of it is decisions related to life and death. It is a widely discussed topic, for example, in the context of ethical issues of accidents involving autonomous cars (see, e.g., Nyholm 2018; Mamak and Glanc 2022). The other domain in which death decisions are widely discussed is the military, where such decisions are embedded in the nature of armed conflicts (see, e.g., Sparrow 2007). In response to the emergence of a deployment system that could make such decisions, a global movement was initiated called "Campaign to Stop Killer Robots" (see, e.g., Solovyeva and Hynek 2023). So, there is a formulated expectation that it should not be possible for machines to make some human decisions or, put differently, human-like robots should not be able to make some decisions.

Staying on the topic of robots in the military context and changing the discussion from "inside" aspects of human likeness to the "external" features, in the paper that Kowalczewska and I wrote, we call for an end to building robots resembling humans that would be deployed in battlefield (Mamak and Kowalczewska 2023). We argue there that human form in the design of robot soldiers could cause unnecessary costs in human life on the other side of the conflict, among the uninvolved actors, and on the side that deployed robots.

The other kinds of human-like robots that are discussed in the context of a potential ban are sex robots. There is a movement that was modeled on the "Campaign to Stop Killer Robots," which is the "Campaign Against Porn Robots" (formerly known as "Campaign Against Sex Robots") (Richardson 2016). In the case of this movement, the aim is clearly stated in its title. It seems, however, that in scholarly works, there is less sympathy to campaign against sex robots than against killer robots (see, e.g., Danaher et al. 2017; Hancock 2020). Disagreement with the idea that all sex robots should be banned does not mean that one cannot oppose some specific kinds of sex robots. The two kinds of such robots that are the focus of most of the discussions in literature are child-like sex robots and rape robots, which are robots that are built with the scenario of rape. For example, Danaher believes that robots of this kind should be prohibited (Danaher 2017). However, some believe that child-like sex robots and rape robots could be useful to some extent, and thus they should not be banned, but instead limited (see, e.g., Levy 2008). The commonest argument in favor of deployment of such robots is the possibility of decreasing the harm done to actual humans. It is based on the hope that people with sexual tendencies that cannot be satisfied without committing a crime could use robots instead of harming people. The potential usefulness of child-like sex robots and rape robots is an empirical matter. Initial data from Italy suggest that sex offenders are less open to sexual relations with sex robots than non-offenders (Zara et al. 2022). Danaher raises doubts about whether there should be experimentation at all in the context of the potential positive impact of discussed categories of sex robots. He points out that research cannot give us decisive data in the foreseeable future, and he maintains the position that responses to such robots should be restricted (Danaher 2019a). Staying on the topic of sex robots, Lancaster argues that robots of this kind should not look like particular people unless it is made with their consent; she calls for criminalization in that respect (Lancaster 2021).

The last topic in this part is related to representational harms. Human-like robots could strengthen or create stereotypes against humans that robots represent. To some extent, it seems that social categorization is an inseparable element of human-like form in robots (such as gender or race). To avoid some of the negative aspects of stereotypization against humans, human form should be abandoned by robots that are meant to perform tasks that in a given society are deemed to be low-esteemed jobs.

Some Aspects of Human Likeness Should be Limited

In the previous section, I focused on specific robots or aspects of human likeness, which are presented with doubts about whether they should exist at all. In this part, I am listing aspects of human likeness in robots that I believe should not be banned as such, but rather should be used in a limited manner and with caution. In general, this book conveys the message that human likeness in robots is inherently dangerous, but here I want to refer to specific aspects of human likeness within the framework presented at the beginning of this chapter.

The last issue that was mentioned in the previous subsection was related to representational aspects of human-like robots. It was mentioned that human likeness should be avoided in the case of robots that are used to perform tasks that could be seen as degrading (if performed by humans) because it could indirectly target the social groups whose traits could be derived from such robots. Here, it can be said that even if it is not planned to use human-like robots to perform tasks that might embarrass some social groups, the social group traits in designing robots should still be used with caution.

The mere presence of human-shaped objects in public spaces is dangerous for humans due to the risk of mistaking them for humans (epistemological threat). Based on that, it might be better to limit the presence of human-like robots in contexts in which those risks are more serious (e.g., in traffic and in public spaces). The limitation of public presence of robots is also relevant from the perspective of indirect risks, such as callousness to the potential suffering of humans. The growing presence of human-shaped objects around us might make us ignore the signs of problems in real humans.

Humans tend to attribute human-like qualities to robots. Because that tendency might increase with greater human likeness, it might not be the best idea to use human-like form in contexts in which robots might not meet the expectations of human likeness. Seeing human-like qualities in them could be dangerous to humans, or harmful in other ways. For example, humans might assume that the human-looking robot security guard could do more than it actually can (less than humans in the same role), giving humans an untrue vision of the level of security.

A potential matter for limitation could be human-like robots that mimic human-like behaviors, for example, deceiving or showing emotional bonds with humans. One of the phenomena related to the discussed problem is love in robots. Sullins noted that mimicking love is easier than mimicking other states because humans want to be loved (Sullins 2012). Nyholm and Frank point out that while designing robots that can say that they love humans, there should be concerns about ethics (Nyholm and Frank 2019).

Contracting Ethical Risks Related to the Deployment of Humans by Accompanying Changes

The last group of inputs that can be drawn from the previous chapters is that the deployment of human-like robots might require some accompanying changes that will counterbalance the disruptive character of human likeness in robots. The focus of the previous subsections was on the negatives (what we should not do), but here the focus is on the positives (what we should do). Changes might be required at the individual level, in the design of robots, and at different levels of regulations.

To avoid having human-like robots that can be mistaken for humans, the roboticness of entities should be observable to humans. This issue is related to the concept of transparency. Transparency in robots can be understood as a lack of deception; the end-user should be kept aware of the robotic nature of the entity (see, e.g., Theodorou et al. 2017:

232). Transparency should be understood extremely broadly. As was discussed in the context of safety problems, mistakes could also occur if the creator has no intention of hiding the robots' real nature. Mistaking humans for robots could also occur in the case of robots that have scant resemblance to humans, for example, in the traffic context. Elsewhere, I noted that sometimes robots should *actively* expose their nature, with warning signals, to make sure that they will not be mistaken for humans even from a distance (Mamak 2024: 112). Robots should also "inform" the other elements of the technological environment that they are robots and not humans to avoid their classification as humans, which could cause problems (Mamak 2024: 112). In that sense, transparency should concern not only humans who could be mistaken but also the whole infrastructure. For example, if a security camera recognizes a fallen humanoid robot as a human and initiates a rescue, then the ambulance that is sent would generate unnecessary costs or, worse, block saving real humans in real danger because there might be no additional ambulance available at the same time. I also proposed that humanoid robots should be obliged by design to take active measures to protect humans, inform them about the dangers by using their sensors, or sacrifice themselves to protect humans (Mamak 2024: 113).

Being uncertain about the ontological status of the human-like shaped entity could create risks for humans. To reduce them, it might be useful to follow Danaher's proposition to treat robots that look like humans as humans in everyday situations (Danaher 2020). We might be compelled by that to treat ethically entities that do not "deserve" such treatment due to their ontological qualities, but at the same time, this attitude gives us more chances to treat humans ethically when they might be mistaken for robots. In other words, when we feel obliged to treat humans and human-like robots in the same way, then human safety is preserved to a greater extent than in a situation in which the norm is to treat the two entities differently.

Mistreating human-like robots could be considered by observers to be an attack on humans. Even if the attacker knows that they are attacking a robot, external evaluation of the event might cause an observer to have an unpleasant feeling, such as observing people being armed. Darling proposes extending legal protection to social robots (Darling 2016), not for the robots themselves but for humans who might be witnessing troublesome spectacles. Elsewhere, I noted that mistreating robots could be considered to be an act against public morality, and I called for a ban on public violence against robots (Mamak 2022b).

The human form in robots might enable the emergence of human-like relations between robots and humans. For example, it is discussed in the context of love and friendship (Danaher 2019b; Sætra 2021). If humans develop such relationships with robots, it might be morally expected to respect such relations (see, e.g., Coeckelbergh 2021; Gunkel 2018b). Elsewhere, I argue that love-like relationships should be protected by criminal law (Mamak 2022a, 2023; see also Sweeney 2023). The rationale is not to protect the robots themselves, but to acknowledge love as a relationship that is important from both moral and legal perspectives, and to protect humans who are in such relationships.

The emergence of human-like robots in the non-superficial, external sense might also require considering robots' interests, treating them in a way that does not cause harm, and sharing their space and resources. In general, this could impose burdens on indicators and society, giving them rights (see, e.g., Hildt 2023; Bryson 2018; Gunkel 2018a; Mamak forthcoming; Friedman 2023).

It was mentioned that humans might include robots in responsibility practices, to some extent treating robots as being responsible for "their" own actions. This tendency creates the risk that real wrongdoers could use robots to limit their own responsibility. To clarify the situation, there is a need to regulate the issue of responsibility, which hinders the use of such a tendency (Feier et al. 2022).

Embedding clues in robots that allow us to see specific social categorization was previously presented as something that might be good to avoid. At the same time, social categorization seems to be unavoidable, to some extent. If people will socially categorize human-like robots anyway, or some robots need to look like humans, mechanisms might be used actively to challenge harmful stereotypes (see, e.g., Eyssel and Hegel 2012; Weßel et al. 2023).

Concerns were formulated that interactions with human-like robots might change the way in which we interact with real humans. In that respect, there are ideas about implementing robot mechanisms that aim to limit their negative effects on humans. Cappuccio et al. argue that robots should be designed in a way that makes humans better people (Cappuccio et al. 2021). However, the agenda of using interaction with robots to improve interactions with humans has a downside. It might be considered as a form of paternalism. As a potential problem for robots, Coggins and Steinert give examples of robots that are programmed not to respond if human queries to them contain swear words (Coggins and

Steinert 2023). A different problem is whether the anticipated effect of the design will actually be achievable. To illustrate that problem, I refer to the idea of embedding in sex robots the consent module that aims to teach users that sex is not always available on demand, and the natural thing is that sometimes the other side could say no; in such circumstances, the refusal needs to be accepted (Peeters and Haselager 2021). In my previous book, I noted that the module, despite noble intentions, might expose rape options to people who did not think about that before, and it could become a hidden rape module with possible negative consequences (Mamak 2023: 82).

If human-like robots are expected to behave like humans, then they should be made in ways that meet those expectations. One of the aspects of this was making robots behave ethically. However, it was mentioned how difficult this agenda is, and some believe that robots will never be human-like in any sense.

From Ethics to Practice

In the preceding discussion, problems with human likeness in robot design were listed, and in some cases, solutions were indicated (such as a moratorium, criminalization, and design choices). Here, the focus is on the means by which problems can be solved or mitigated. There is an overview of selected views that are formulated while discussing the ways to ensure that robots will be beneficial. The aim of this selection is to show the multidimensionality of this task.

Thinking about responses to ethical issues is complicated because the problems with human-like robots are related not only to the technical characteristics of robots but also to the ways in which human-like robots interact with humans. Weng points out at the "robot sociability problem" to name the issues that arise when robots are integrated into human society (Weng 2010). Those are not issues that are immediately apparent when introducing robots; the way in which robots will be embedded in social life is to some extent independent of the intentions of their creators. Nevertheless, while talking about regulations, all aspects related to human-like robots might need to be considered.

Regulation should be understood broadly, following the explanation presented by Leenes et al.: "The regulation is aimed at influencing the behavior of people in the context of developments in the field of robotics" (Leenes et al. 2017: 6). They underline that the law is the most obvious example of regulation, but the regulation should be understood broader, to cover such things as social norms, market, and the architecture (the technology itself).

One example of regulation in a broad sense is the code of ethics. Riek and Howard propose such a solution for the field of human–robot interactions that covers many of the issues discussed before. They start with the "prime directive," which they explain is respect for a person, which includes respect for autonomy and bodily and mental integrity. They add here that robots should behave in the way in which we would expect respectful humans to behave. Besides that guiding principle, they propose more specific ones that are divided into four categories: human dignity (i.e., need for respecting emotional needs and psychological and physical needs), design considerations (transparency, predictability, trustworthiness), legal (privacy issues, respecting human laws and regulation, the capacity to track the issues for litigation), and social considerations (problem of attachments with robots, morphology related to racist or sexist design choices). Some of the design considerations are transparency, predictability, or the so-called kill switch. In summary, they underline that "human morphology and functionality is permitted only to the extent necessary for the achievement of reasonable design objectives" (Riek and Howard 2014: 6). The code of ethics is one of many regulatory means, but more things should be considered.

Veruggio et al. discuss how ethical deliberation regarding robots could be incorporated into laws and rights. The authors mention many forms of expression related to ethical issues that have the potential to transform the world of robotics into one that is more mingled with ethical deliberations, such as oaths (the Hippocratic oath), manifestos, declarations, resolutions, more technical standards (ISO), ending with national laws (Veruggio et al. 2016: 2156). Moon et al. focus on the ethics of the embodied AIs (corporal) and how to bridge ethical thinking into robot engineering design. They mention technology governance and the coordinated research that aim to develop robots with ethics in mind (Moon et al. 2021: 227). Weng focuses on entities that should be involved in the process and mentions that citizens, parliaments, and agencies should adopt the basic principles regarding AI and robotics in their related work to ensure that the robots will (Weng 2024: 448).

Shneiderman, in his book about human-centered AI, which also covers the problem of robots, describes four levels of governance structures that include the engagement of various stakeholders who have an impact on the final product: software engineering practices (within teams), safety culture (within organizations), trustworthy certification by independent organizations (by industry-specific organizations, audit and insurance companies, research institutes, civil society organizations), and finally, regulation by governmental agencies (Shneiderman and Shneiderman 2022: 141–42). In his book *We, the Robots?* Chesterman underlines that AI poses difficulties for regulation to an extent that it should not be left to national countries alone. Considering the speed of innovation, regulatory efforts should also be made at the international level. He suggests that the nature of risks that AI poses to the legitimacy of public authority should be considered (Chesterman 2021).

Pasquale, in his book *New Laws of Robotics,* underlines the role of humans in the process of regulation of robots to ensure that robots will work for us and not otherwise. The proposed laws are: first, robotic systems and AI should complement professionals, not replace them; second, robotic systems and AI should not counterfeit humanity; third, robotic systems and AI should not intensify zero-sum arms races; fourth, robotic systems and AI must always indicate the identity of their creator(s), controller(s), and owner(s) (Pasquale 2020: 3–12). The second and fourth laws are especially important from the perspective of this book. When the second law is discussed, Pasquale calls for minimalization of the use of human form in robots. Discussing the fourth law, he is referring to the problem of responsibility; calling for robots to be built in ways that ensure that it will be possible to hold someone responsible, he mentions concepts like "responsibility by design", "security-by-design," and "privacy-by-design" (Pasquale 2020: 12).

The calls for regulation have been made in recent years, and actual (partial) regulations have been accompanied at different levels. There is a regulatory boom related to the development of AI. For example, numerous ethical guidelines have been published (see, e.g., Hagendorff 2023). What is more, countries or groups of countries, like in the European Union, work on their own laws (see, e.g., Ulnicane 2022). Langman et al. point out that most attention so far has been given to the regulation of AI instead of robots; they guess that it is because AI will be incorporated into all aspects of life (Langman et al. 2021: 6). Some of the issues covered during discussion on AI also concern robotics, but there may be a need to have more detailed regulations that are related to the embodied character of robotics (cf. Pagallo 2013; Balkin 2015; Calo 2015; Turner 2019; Abbott 2020; Gellers 2020; Darling 2021; Hallevy 2013).

To conclude this section, several ways of thinking were presented about the ways in which ethical deliberation regarding human-like robots could be transferred into practice, from the change of social norms to hard regulation at the national and international levels. Several entities need to be engaged with, and a range of means need to be adopted. Without a doubt, the problem of making human-like robots beneficial to humans is complex. Karnouskos recognizes the complexity of endeavoring to regulate, and he underlines the need for interdisciplinary thinking in the discussion on the regulation of robots. He notes that law and society as a whole are not yet prepared for the prevalence of robots (Karnouskos 2022). Coeckelbergh also points out that in the policy discussion about AI, there is a need for interdisciplinarity and transdisciplinarity to take into account the humanities and social sciences (Coeckelbergh 2020: 177–79).

In the next section, I focus on roboactivism to underline the need for proactive involvement by people who have a smaller or bigger impact on what robots are around us.

Roboactivism

"Roboactivism" is the idea based on merging two concepts—"robot" and "activism"—like robotics that combines ethics and robots (Veruggio 2005) and robophilsophy that refers to the philosophical issues around robotics (see Hakli et al. 2023). According to the *Oxford English Dictionary*, "activism" may be understood as: "The policy of active participation or engagement in a particular sphere of activity; spec. the use of vigorous campaigning to bring about political or social change." In the context of robots, roboactivism means activism around the deployment of robots to make sure that robots are beneficial to society. Here, I endorse the position of Green, who said that data scientists could not be neutral while doing their job:

"I argue that data science must embrace a political orientation. Data scientists must recognize themselves as political actors engaged in normative constructions of society and evaluate their work according to its downstream impacts on people's lives" (Green 2021: 249).

The people responsible for the design of robots or their deployment should recognize their role and take an active part in ensuring that robots will benefit society. Who are those people? All who may have the power to have an impact somehow on the final "look" of robots operating among other designers, companies producing robots, sellers, buyers (e.g., the police, healthcare), policymakers, and academics.

At the abstract level, it might sound good, but there are potential problems with putting the idea into practice. As was shown before, there is already a discussion on robots with opposing positions that could not agree on which specific robots should be around us and which should not. I already mentioned that the "Campaign Against Sex Robots" (now the "Campaign Against Porn Robots") (cf. Richardson 2016; Richardson and Odlind 2023) is concerned about the impact of human-like sex robots on women. The aim of this campaign is to ban sex robots but it is met with opposition that underlines the positive aspects of sex robots (Danaher et al. 2017). So, there could be problems with recognizing what is good and what is wrong, and in which direction we should go with robots. The other issue is that there could be agreement on the direction, but there could be disagreement about the means and their effects. For example, agreeing that robots need to be more diverse is one thing, and the other is how to achieve that, Williams is concerned that noble intentions could bring negative consequences (Williams 2023).

Nevertheless, acknowledging the limitations of roboactivism, I think that it is useful to recognize that the people engaging in the robotic project could impact the world, their decisions could shape the ethical landscape, and, to some extent, they have the power to impact that landscape, to complement the attempts of making robots beneficial to humans.

Conclusions

This concluding chapter organizes the specific inputs formulated earlier in the book. The book's general tone is skeptical about the idea of creating human-like robots. However, even if there are arguments against the whole project, from a practical point of view, the project seems to be unstoppable. Such robots are being created and will be created in the future. To ensure that human-like robot projects will be ethical, certain aspects of human-like robots should be banned or limited. There is also a need to make some positive changes to counter any negative consequences of the presence of human-like robots. This chapter also discusses how ethical deliberation could be transferred into practice, underlining the multidimensionality of the problem. The broadly understood regulatory activities need to concern the different activities of different actors, starting with individuals and ending with countries and international organizations. There is also a formulated call for roboactivism, taking positive actions to shape the reality of human-like robots in line with human values by all the people engaged in the design and deployment of robots.

References

Abbott, Ryan. 2020. *The Reasonable Robot: Artificial Intelligence and the Law*. Cambridge: Cambridge University Press. https://doi.org/10.1017/9781108631761.

Balkin, Jack. 2015. "The Path of Robotics Law." *California Law Review* 6. https://digitalcommons.law.yale.edu/fss_papers/5150.

Bendel, Oliver. 2021. "Love Dolls and Sex Robots in Unproven and Unexplored Fields of Application." *Paladyn, Journal of Behavioral Robotics* 12 (1): 1–12. https://doi.org/10.1515/pjbr-2021-0004.

Bryson, Joanna J. 2018. "Patiency Is Not a Virtue: The Design of Intelligent Systems and Systems of Ethics." *Ethics and Information Technology* 20 (1): 15–26. https://doi.org/10.1007/s10676-018-9448-6.

Calo, Ryan. 2015. "Robotics and the Lessons of Cyberlaw." *California Law Review* 103 (January): 513.

Cappuccio, Massimiliano L., Eduardo B. Sandoval, Omar Mubin, Mohammad Obaid, and Mari Velonaki. 2021. "Can Robots Make Us Better Humans?" *International Journal of Social Robotics* 13 (1): 7–22. https://doi.org/10.1007/s12369-020-00700-6.

Chesterman, Simon. 2021. *We, the Robots? Regulating Artificial Intelligence and the Limits of the Law*. Cambridge University Press.

Chomanski, Bartek. 2019. "What's Wrong with Designing People to Serve?" *Ethical Theory and Moral Practice* 22 (4): 993–1015. https://doi.org/10.1007/s10677-019-10029-3.

Coeckelbergh, Mark. 2020. *AI Ethics*. Cambridge, MA: MIT Press. https://mitpress.mit.edu/books/ai-ethics.

———. 2021. "Should We Treat Teddy Bear 2.0 as a Kantian Dog? Four Arguments for the Indirect Moral Standing of Personal Social Robots, with Implications for Thinking about Animals and Humans." *Minds and Machines* 31 (3): 337–60. https://doi.org/10.1007/s11023-020-09554-3.

Coggins, Tom N., and Steffen Steinert. 2023. "The Seven Troubles with Norm-Compliant Robots." *Ethics and Information Technology* 25 (2): 29. https://doi.org/10.1007/s10676-023-09701-1.

Danaher, John. 2017. "Robotic Rape and Robotic Child Sexual Abuse: Should They Be Criminalised?" *Criminal Law and Philosophy* 11 (1): 71–95. https://doi.org/10.1007/s11572-014-9362-x.

———. 2019a. "Regulating Child Sex Robots: Restriction or Experimentation?" *Medical Law Review* 27 (4): 553–75. https://doi.org/10.1093/medlaw/fwz002.

———. 2019b. "The Philosophical Case for Robot Friendship." *Journal of Posthuman Studies*.

———. 2020. "Welcoming Robots into the Moral Circle: A Defence of Ethical Behaviourism." *Science and Engineering Ethics* 26: 2023–49. https://doi.org/10.1007/s11948-019-00119-x.

Danaher, John, Brian Earp, and Anders Sandberg. 2017. "Should We Campaign against Sex Robots?" In *Robot Sex: Social and Ethical Implications*, edited by John Danaher and Neil McArthur, 47–72. Cambridge, MA: MIT Press. https://doi.org/10.7551/mitpress/9780262036689.003.0004.

Darling, Kate. 2016. "Extending Legal Protection to Social Robots: The Effects of Anthropomorphism, Empathy, and Violent Behavior towards Robotic Objects." In *Robot Law*, edited by Ryan Calo, A. Froomkin, and Ian Kerr, 213–32. Edward Elgar Publishing. https://doi.org/10.4337/9781783476732.00017.

———. 2021. *The New Breed: How to Think about Robots*. Dublin: Allen Lane. https://us.macmillan.com/thenewbreed/katedarling/9781250296115.

Döring, Nicola, M. Rohangis Mohseni, and Roberto Walter. 2020. "Design, Use, and Effects of Sex Dolls and Sex Robots: Scoping Review." *Journal of Medical Internet Research* 22 (7): e18551. https://doi.org/10.2196/18551.

Eyssel, Friederike, and Frank Hegel. 2012. "(S)He's Got the Look: Gender Stereotyping of Robots." *Journal of Applied Social Psychology* 42 (9): 2213–30. https://doi.org/10.1111/j.1559-1816.2012.00937.x.

Feier, Till, Jan Gogoll, and Matthias Uhl. 2022. "Hiding Behind Machines: Artificial Agents May Help to Evade Punishment." *Science and Engineering Ethics* 28 (2): 19. https://doi.org/10.1007/s11948-022-00372-7.

Fosch-Villaronga, Eduard, and Adam Poulsen. 2020. "Sex Care Robots: Exploring the Potential Use of Sexual Robot Technologies for Disabled and Elder Care." *Paladyn, Journal of Behavioral Robotics* 11 (1): 1–18. https://doi.org/10.1515/pjbr-2020-0001.

Friedman, Cindy. 2023. "Granting Negative Rights to Humanoid Robots." *Social Robots in Social Institutions*, 145–54. https://doi.org/10.3233/FAIA220613.

Gellers, Joshua C. 2020. *Rights for Robots: Artificial Intelligence, Animal and Environmental Law*. London: Routledge. https://doi.org/10.4324/9780429288159.

Green, Ben. 2021. "Data Science as Political Action: Grounding Data Science in a Politics of Justice." *Journal of Social Computing* 2 (3): 249–65. https://doi.org/10.23919/JSC.2021.0029.

Gunkel, David J. 2018a. *Robot Rights*. Cambridge, MA: MIT Press.

———. 2018b. "The Other Question: Can and Should Robots Have Rights?" *Ethics and Information Technology* 20 (2): 87–99. https://doi.org/10.1007/s10676-017-9442-4.

Hagendorff, Thilo. 2023. "AI Ethics and Its Pitfalls: Not Living up to Its Own Standards?" *AI and Ethics* 3 (1): 329–36. https://doi.org/10.1007/s43681-022-00173-5.

Hakli, Raul, Pekka Mäkelä, and Johanna Seibt, eds. 2023. *Social Robots in Social Institutions: Proceedings of Robophilosophy 2022*. IOS Press.

Hallevy, Gabriel. 2013. *When Robots Kill: Artificial Intelligence under Criminal Law*. Boston: Northeastern.

Hancock, Eleanor. 2020. "Should Society Accept Sex Robots? Changing My Perspective on Sex Robots through Researching the Future of Intimacy." *Paladyn, Journal of Behavioral Robotics* 11 (1): 428–42. https://doi.org/10.1515/pjbr-2020-0025.

Hildt, Elisabeth. 2023. "The Prospects of Artificial Consciousness: Ethical Dimensions and Concerns." *AJOB Neuroscience* 14 (2): 58–71. https://doi.org/10.1080/21507740.2022.2148773.

Karnouskos, Stamatis. 2022. "Symbiosis with Artificial Intelligence via the Prism of Law, Robots, and Society." *Artificial Intelligence and Law* 30 (1): 93–115. https://doi.org/10.1007/s10506-021-09289-1.

Lancaster, Karen. 2021. "Non-consensual Personified Sexbots: An Intrinsic Wrong." *Ethics and Information Technology* 23 (4): 589–600. https://doi.org/10.1007/s10676-021-09597-9.

Langman, Sofya, Nicole Capicotto, Yaser Maddahi, and Kourosh Zareinia. 2021. "Roboethics Principles and Policies in Europe and North America." *SN Applied Sciences* 3 (12): 857. https://doi.org/10.1007/s42452-021-04853-5.

Leenes, Ronald, Erica Palmerini, Bert-Jaap Koops, Andrea Bertolini, Pericle Salvini, and Federica Lucivero. 2017. "Regulatory Challenges of Robotics: Some Guidelines for Addressing Legal and Ethical Issues." *Law, Innovation and Technology* 9 (1): 1–44. https://doi.org/10.1080/17579961.2017.1304921.

Levy, David. 2008. *Love and Sex with Robots: The Evolution of Human–Robot Relationships*. New York: Harper Perennial.

Mamak, Kamil. 2021. "Whether to Save a Robot or a Human: On the Ethical and Legal Limits of Protections for Robots." *Frontiers in Robotics and AI* 8. https://doi.org/10.3389/frobt.2021.712427.

———. 2022a. "Should Criminal Law Protect Love Relation with Robots?" *AI & SOCIETY*, April. https://doi.org/10.1007/s00146-022-01439-6.

———. 2022b. "Should Violence against Robots Be Banned?" *International Journal of Social Robotics* 14 (4): 1057–66. https://doi.org/10.1007/s12369-021-00852-z.

———. 2023. *Robotics, AI and Criminal Law: Crimes against Robots*. 1st edition. London: Routledge. https://doi.org/10.4324/9781003331100.

———. 2024. "Challenges of the Legal Protection of Human Lives in the Time of Anthropomorphic Robots." In *Cambridge Handbook on Law, Policy, and Regulations for Human–Robot Interaction*, edited by Woodrow Barfield, Yueh-Hsuan Weng, and Ugo Pagallo. Cambridge University Press.

———. forthcoming. "Toward a Universal Declaration of Robot Rights? Building Robots into Global Diversity." In *Handbook of Research on Global Diversity Management*, edited by Mustafa F. Ozbilgin and Cihat Erbil. Edward Elgar.

Mamak, Kamil, and Jadwiga Glanc. 2022. "Problems with the Prospective Connected Autonomous Vehicles Regulation: Finding a Fair Balance versus the Instinct for Self-Preservation." *Technology in Society*, September, 102127. https://doi.org/10.1016/j.techsoc.2022.102127.

Mamak, Kamil, and Kaja Kowalczewska. 2023. "Military Robots Should Not Look Like a Humans." *Ethics and Information Technology* 25. https://doi.org/10.1007/s10676-023-09718-6.

McArthur, Neil. 2017. "The Case for Sexbots." In *Robot Sex: Social and Ethical Implications*, edited by John Danaher and Neil McArthur, 31–46. Cambridge, MA: MIT Press. https://doi.org/10.7551/mitpress/9780262036689.003.0003.

Metzinger, Thomas. 2021. "Artificial Suffering: An Argument for a Global Moratorium on Synthetic Phenomenology." *Journal of Artificial Intelligence and Consciousness* 8 (1): 43–66. https://doi.org/10.1142/S270507852150003X.

Moon, AJung, Shalaleh Rismani, and H. F. Machie Van der Loos. 2021. "Ethics of Corporeal, Co-present Robots as Agents of Influence: A Review." *Current Robotics Reports* 2: 223–29. https://doi.org/10.1007/s43154-021-00053-6

Müller, Vincent C. 2014. "Risks of General Artificial Intelligence." *Journal of Experimental & Theoretical Artificial Intelligence* 26 (3): 297–301. https://doi.org/10.1080/0952813X.2014.895110.

Musiał, Maciej. 2017. "Designing (Artificial) People to Serve—The Other Side of the Coin." *Journal of Experimental & Theoretical Artificial Intelligence* 29 (5): 1087–97. https://doi.org/10.1080/0952813X.2017.1309691.

———. 2022. "Can We Design Artificial Persons without Being Manipulative?" *AI & SOCIETY*, October. https://doi.org/10.1007/s00146-022-01575-z.

Nyholm, Sven. 2018. "The Ethics of Crashes with Self-Driving Cars: A Roadmap, I." *Philosophy Compass* 13 (7): e12507. https://doi.org/10.1111/phc3.12507.

Nyholm, Sven, and Lily Eva Frank. 2019. "It Loves Me, It Loves Me Not: Is It Morally Problematic to Design Sex Robots That Appear to Love Their Owners?" *Techné: Research in Philosophy and Technology* 23 (3): 402–24. https://doi.org/10.5840/techne2019122110.

Ord, Toby. 2020. *The Precipice: Existential Risk and the Future of Humanity.* Illustrated edition. New York: Hachette Books.

Pagallo, Ugo. 2013. *The Laws of Robots.* Dordrecht: Springer Netherlands. https://doi.org/10.1007/978-94-007-6564-1.

Pasquale, Frank. 2020. *New Laws of Robotics: Defending Human Expertise in the Age of AI.* Cambridge, MA: The Belknap Press of Harvard University Press. https://www.hup.harvard.edu/catalog.php?isbn=9780674975224.

Peeters, Anco, and Pim Haselager. 2021. "Designing Virtuous Sex Robots." *International Journal of Social Robotics* 13 (1): 55–66. https://doi.org/10.1007/s12369-019-00592-1.

Petersen, Steve. 2011. "Designing People to Serve." In *Robot Ethics*, edited by Patrick Lin, George Bekey, and Keith Abney. Cambridge, MA: MIT Press. https://philarchive.org/rec/PETDPT.

Richardson, Kathleen. 2016. "Are Sex Robots as Bad as Killing Robots?" *What Social Robots Can and Should Do*, 27–31. https://doi.org/10.3233/978-1-61499-708-5-27.

Richardson, Kathleen, and Charlotta Odlind, eds. 2023. *Man-made Women: The Sexual Politics of Sex Dolls and Sex Robots.* Springer Nature.

Riek, Laurel, and Don Howard. 2014. "A Code of Ethics for the Human–Robot Interaction Profession." SSRN Scholarly Paper. Rochester, NY. https://papers.ssrn.com/abstract=2757805.

Russell, Stuart. 2019. *Human Compatible: AI and the Problem of Control.* Allen Lane.

Sætra, Henrik Skaug. 2021. "Loving Robots Changing Love: Towards a Practical Deficiency-Love." *Journal of Future Robot Life* (Preprint): 1–19. https://doi.org/10.3233/FRL-200023.

Schwitzgebel, Eric, and Mara Garza. 2020. "Designing AI with Rights, Consciousness, Self-Respect, and Freedom." In *Ethics of Artificial Intelligence*, edited by S. Matthew Liao, 459–79. New York: Oxford University Press.

Shneiderman, Ben, and Ben Shneiderman. 2022. *Human-centered AI*. Oxford, New York: Oxford University Press.

Solovyeva, Anzhelika, and Nik Hynek. 2023. "When Stigmatization Does Not Work: Over-securitization in Efforts of the Campaign to Stop Killer Robots." *AI & SOCIETY* 38 (6): 2547–69. https://doi.org/10.1007/s00146-022-01613-w.

Sparrow, Robert. 2007. "Killer Robots." *Journal of Applied Philosophy* 24 (1): 62–77.

———. 2020. "Robotics Has a Race Problem." *Science, Technology, & Human Values* 45 (3): 538–60. https://doi.org/10.1177/0162243919862862.

Sullins, John P. 2012. "Robots, Love, and Sex: The Ethics of Building a Love Machine." *IEEE Transactions on Affective Computing* 3 (4): 398–409. https://doi.org/10.1109/T-AFFC.2012.31.

Sweeney, Paula. 2023. "Could the Destruction of a Beloved Robot Be Considered a Hate Crime? An Exploration of the Legal and Social Significance of Robot Love." *AI & SOCIETY*, November. https://doi.org/10.1007/s00146-023-01805-y.

Theodorou, Andreas, Robert H. Wortham, and Joanna J. Bryson. 2017. "Designing and Implementing Transparency for Real Time Inspection of Autonomous Robots." *Connection Science* 29 (3): 230–41. https://doi.org/10.1080/09540091.2017.1310182.

Turner, Jacob. 2019. *Robot Rules: Regulating Artificial Intelligence*. Cham: Springer International Publishing. https://doi.org/10.1007/978-3-319-96235-1.

Ulnicane, Inga. 2022. "Artificial Intelligence in the European Union." In *The Routledge Handbook of European Integrations*, edited by Thomas Hoerber, Gabriel Weber, and Ignazio Cabras. Taylor & Francis. https://doi.org/10.4324/9780429262081-19.

Veruggio, Gianmarco. 2005. "The Birth of Roboethics." In *ICRA 2005, IEEE International Conference on Robotics and Automation, Workshop on Roboethics*, edited by John Doe, 1–4. https://philarchive.org/rec/VERTBO-3.

Veruggio, Gianmarco, Fiorella Operto, and George Bekey. 2016. "Roboethics: Social and Ethical Implications." In *Springer Handbook of Robotics*, edited by Bruno Siciliano and Oussama Khatib, 2135–60. Springer Handbooks. Cham: Springer International Publishing. https://doi.org/10.1007/978-3-319-32552-1_80.

Vold, Karina, and Daniel R. Harris. 2021. "How Does Artificial Intelligence Pose an Existential Risk?" In *The Oxford Handbook of Digital Ethics*, edited by Carissa Véliz, 724–47. https://doi.org/10.1093/oxfordhb/9780198857815.013.36.

Weng, Yueh-Hsuan. 2010. "Beyond Robot Ethics: On a Legislative Consortium for Social Robotics." *Advanced Robotics* 24 (13): 1919–26. https://doi.org/10.1163/016918610X527220.

———. 2024. "Ethical Design and Standardization for Robot Governance." In *Cambridge Handbook on Law, Policy, and Regulations for Human–Robot Interaction*, edited by Woodrow Barfield, Yueh-Hsuan Weng, and Ugo Pagallo. Cambridge University Press.

132 | *Ethics in Human-like Robots*

Weßel, Merle, Niklas Ellerich-Groppe, and Mark Schweda. 2023. "Gender Stereotyping of Robotic Systems in Eldercare: An Exploratory Analysis of Ethical Problems and Possible Solutions." *International Journal of Social Robotics* 15 (11): 1963–76. https://doi.org/10.1007/s12369-021-00854-x.

Williams, Tom. 2023. *The Eye of the Robot Beholder: Ethical Risks of Representation, Recognition, and Reasoning over Identity Characteristics in Human–Robot Interaction.* In *Companion of the 2023 ACM/IEEE International Conference on Human–Robot Interaction,* March, HRI'23, 1–10. https://doi.org/10.1145/3568294.3580031.

Zara, Georgia, Sara Veggi, and David P. Farrington. 2022. "Sexbots as Synthetic Companions: Comparing Attitudes of Official Sex Offenders and Non-Offenders." *International Journal of Social Robotics* 14 (2): 479–98. https://doi.org/10.1007/s12369-021-00797-3.

Index